SOBER TRAVEL HANDBOOK

NAVIGATING THE WORLD ALCOHOL-FREE

TERESA BERGEN

Portland, OR | Cleveland, OH

Sober Travel Handbook: Navigating the World Alcohol-Free
© Teresa Bergen, 2026
First edition © Microcosm Publishing, 2026
Book design by Joe Biel and Sarah Koch
Edited by Katie Haegele
Cover photo by Lies Thru a Lens
ISBN 9781648414633
This is Microcosm #954

For a catalog, write or visit:
Microcosm Publishing
2752 N Williams Ave.
Portland, OR 97227
(503)799-2698

All the news from the misfits in print at Microcosm.Pub/Newsletter
Get more copies of this book at Microcosm.Pub/SoberTravelHandbook
Find more books by Teresa Bergen at Microcosm.Pub/TeresaBergen
EU Safety Information: microcosmpublishing.com/gpsr

To join the ranks of high-class stores that feature Microcosm titles, talk to your rep: In the U.S. COMO (Atlantic), ABRAHAM (Midwest), SALES@ MICROCOSM.PUB (Texas, Oklahoma, Arkansas, Louisiana), IMPRINT (Pacific), TURNAROUND (UK/Africa, Europe, Middle East), UTP/MANDA (Canada), NEWSOUTH (Australia/New Zealand), APD (Asia), HarperCollins (India), and FAIRE in the gift trade.

Did you know that you can buy our books directly from us at sliding scale rates? Support a small, independent publisher and pay less than Amazon's price at www. Microcosm.Pub.

Global labor conditions are bad, and our roots in industrial Cleveland in the '70s and '80s made us appreciate the need to treat workers right. Therefore, our books are MADE IN THE USA.

LCCN 2025027792

ABOUT THE PUBLISHER

MICROCOSM · PUBLISHING

MICROCOSM PUBLISHING is Portland's most diversified publishing house and distributor, with a focus on the colorful, authentic, and empowering. Our books and zines have put your power in your hands since 1996, equipping readers to make positive changes in their lives and in the world around them. Microcosm emphasizes skill-building, showing hidden histories, and fostering creativity through challenging conventional publishing wisdom with books and bookettes about DIY skills, food, bicycling, gender, self-care, and social justice. What was once a distro and record label started by Joe Biel in a drafty bedroom was determined to be *Publishers Weekly*'s fastest-growing publisher of 2022 and #3 in 2023, and is now among the oldest independent publishing houses in Portland, OR, and Cleveland, OH. We are a politically moderate, centrist publisher in a world that has inched to the right for the past 80 years.

Global labor conditions are bad, and our roots in industrial Cleveland in the '70s and '80s made us appreciate the need to treat workers right. Therefore, our books are MADE IN THE USA.

CONTENTS

INTRODUCTION

*P*eople who drink normally might be confused by this book. What's the point? You just travel and don't pick up a drink. Simple. But nondrinkers understand that travel has certain difficulties, especially if you don't drink because you have had problems with alcohol. So often, a drink is only an arm's length away. If you're at a restaurant, it might be at your companion's place, just inches from you. If you're staying in a hotel, there could be cute little bottles in a minibar. It would be so easy to reach out and grab one. Or somebody who doesn't know you're a nondrinker might hand you a glass of champagne at an event. It's such a natural reaction to accept what's given.

Then again, maybe you never had a drinking problem, but it seems like every time you go on vacation, all people do is drink. Many countries have high alcohol use, and nondrinkers living in them or visiting them can feel left out, cranky and uncomfortable in booze-obsessed societies.

At home, nondrinkers know where to go and where to avoid. It's different when you travel. New situations and places can take you by surprise. That drink you've warily avoided may be closer than ever, especially if you're traveling for a professional conference, where connections are often made in a hotel bar. Or for a family wedding or funeral, where emotions run high. Or on a cruise, where the smiling crew members walk around, extending big trays of umbrella-decorated drinks in your direction. Or on an airplane, when you get upgraded to first class and liquor is free.

The biggest stroke of luck in my life was getting out from under my addictions at an early age. While I haven't had a totally charmed life, I've had practically a whole adult lifespan to live fully and to expand my world.

I've always loved to both travel and to write. My first travel writing assignment was in the early nineties, when I spent six months traveling in Asia and wrote a series of dispatches about sobriety abroad for a long-defunct newspaper called *Sober Times*. I've done several kinds of writing in the interim, and plenty of travel. For the last 14 years, travel writing has taken me around the world, exposed me to cultures, situations and people that were new to me, and presented me with approximately one million opportunities to dodge drinks. I've traveled to nature-focused places where drinking just doesn't fit, such as following eagle rays while diving in the Maldives and holding on tight so I don't fall into giant rapids while rafting the Grand Canyon. I've been to countries where the norm is not to drink, such as Jordan, and seen fellow tourists forgo the best sunset spot so they could instead sip wine in a hotel bar. I've spent many clearheaded days and nights learning about other cultures, and noting the place of alcohol in each.

In this book, I'll share the ways I've stayed sober, plus tips from other nondrinking travelers and addiction experts. I found these wise people in many ways. Some are longtime friends from my personal life, others I know from professional organizations I belong to. I recruited some through Facebook and LinkedIn, or referrals from friends. Others I tracked down through their websites or official channels, notably George Koob, director of the National Institute on Alcohol Abuse and Alcoholism. I especially appreciated Dr. Koob taking the time to talk to me before I'd even found a publisher. Fortunately, people saw the value of my aim to help people travel sober, and to normalize nondrinking.

Living in the hard-drinking town of New Orleans helped me quit at a very young age. When I'd been sober for about a week, I was walking down a street in the French Quarter when I noticed there were things *inside* the shop windows. Yes, I realize this doesn't sound like much of a revelation. But I'd been stomping through the Quarter for a year with tunnel vision, focused on my next drink. Most of the windows might as well have been walls for the amount of attention I'd paid them. Just like pondering what's beyond the window, every time I travel somewhere new I go beyond a name, a spot on a map, a hazy, half-baked idea about what that place and those people are all about.

If you're reading this because you want to step beyond the familiar but are afraid of drinking, know that it is possible. Go slowly. Stay vigilant. Be gentle with yourself. Little by little, you can expand your world, too.

WHAT IS SOBER TRAVEL?

One of the challenging parts of writing this book is defining the term "sober travel." The simple answer is it means traveling without drinking. When I say this to anyone who works for a restaurant, hotel, or destination marketing organization, they usually say, "Oh, we have wonderful mocktails." I greatly appreciate that the more delicious alternatives there are, the more people will at least sometimes choose those, and nondrinking will become more normalized. But is there something more to sober travel than doing the exact same thing as usual, except replacing regular tequila with zero-proof tequila in your mojito?

Yes, there is more. Just as there is more to life than sitting and drinking in a bar, there is more to travel than drinking a non-alc drink at a resort bar. When we choose not to drink, we are putting ourselves out in the world without a highly popular and relied upon social crutch. As travel writer and editor Elizabeth Harryman Lasley—a nondrinker by choice rather than necessity—said, "I want to be my real self all the time—not some phony self created by an artificial substance."

So how do we go forth unconventionally, without artifice or anesthesia? I was recently inspired by hearing Jeff Jenkins give a talk for the Adventure Travel Trade Association on living boldly. Jenkins is the creator of the website Chubby Diaries and the Nat Geo show *Never Say Never*, which features himself—a large Black man—taking on adventures like rafting down a 25-foot waterfall. In a video clip from his rafting expedition, the audience clearly sees the look of "what the hell am I doing" on Jeff's face just before he plunges down the falls—and joyously lives to tell. As Jeff emphasized in his speech, life begins where your comfort zone ends.

I think this is especially true for people in recovery. Maybe you got extremely comfy on that bar stool in your favorite bar, or at home boozing it up on your couch. Then in recovery, maybe we get on a predictable schedule of going to certain therapy groups and hanging out with other people like us, or staying home and relying on support from the sober cybersphere. Same if you don't drink for religious reasons—you might spend most of your time with others of the same faith, going to the same places to worship and socialize. While it's vital to have some kind of a support group, we can clip our horizons without even noticing. Untended, our comfort zones have a way of shrinking.

Travel does just the opposite. When we go to new places with new people, we find ourselves in situations we never expected. These could be fabulous or trying, but either way our worldview expands. Just like Jenkins never imagined a chubby Black man like himself would go on big adventures, many people in recovery don't expect to chart a course from barstool to life-changing experiences traveling the globe.

So what makes for successful sober travel beyond just not drinking? Incorporating elements of these four long-term indicators of happy sobriety will turn your vacation into something much more meaningful. And while you could say this is good advice for anybody, drinker or not, it's especially important for those who don't drink in an alcohol-fixated society. Focusing on these elements could make the difference between white-knuckling it through a "vacation" dominated by other people's drunken revelry and having a life-enhancing experience away from home.

BENEFITS OF SOBER TRAVEL

Health

Taking time away from your regular routine can let you focus on what you need for your physical and mental health. Many people find they get more exercise while on vacation, especially desk workers. At a beach they may have more opportunities to surf, swim, and snorkel. Or if they're in an unfamiliar city without a car, they'll explore on foot. Then again, maybe you need more rest, meditation, and reflection during your vacation. You might also think more about your diet rather than just eating what's handy.

Connection

People need to connect with each other—and this is done more authentically when your mind isn't veiled by alcohol. If you're traveling with a partner, friends, family, or colleagues, this is a time to connect with them more deeply. You can also meet other travelers or local people at your destination, especially if you're going solo.

Nature lovers like me will also feel a connection to rivers, landscapes, and wildlife. Once I was paddleboarding at Trillium Lake in Oregon when I idly thought, these are funny-looking fish with their little legs. Wait, fish don't have legs! I realized I was surrounded by pairs of newts doing a spiral mating dance in the water. When we travel to awe-inspiring places, we may also feel more connected to a higher power and/or to more profound aspects of ourselves.

Adventure

This is the part where we expand our horizons. "Adventure" is a subjective concept. Maybe you're on an obvious adventure, like rafting down that 25-foot waterfall. Or the adventure could be more modest, like navigating a country where you don't speak the language, camping when you're more experienced at hotel living, or trying foods that seem very strange to you. Adventure is scalable. You don't have to go for the biggest one first. Even baby steps will expand your comfort zone.

But remember: our brains respond to novelty. That's why we're always reading about how learning a new language or taking a dance class can help stave off dementia. The part of the brain called the interpeduncular nucleus (IPN) "is at the heart of a neural circuit that controls a person's interest in new things, leading him or her to explore them further," according to neurobiologist Dr. Andrew Tapper. Research suggests that dysregulation of the IPN circuit could play a role in several disorders, such as addiction, where patients prefer novel sensation seeking.[1] I suggest travel as a way to light up the brain's novelty circuits without taking drugs.

Contribution

If you're a 12-step person, chances are that you've served as a greeter, coffee maker, or meeting secretary. Same if you don't drink for religious reasons—you've probably helped out in your mosque, church, or temple. Contributing connects us to others

---

1 "Researchers Uncover Neural Circuit that Underlies Interest in Novelty: Brain & Behavior Research Foundation," bbrfoundation.org/content/researchers-uncover-neural-circuit-underlies-interest-novelty

and helps us develop a sense of purpose. It helps us value ourselves and gives us self-respect. While traveling, you could contribute in a big way, like going on a voluntourism trip to build a school in an impoverished area. Or you can contribute in a constant stream of little ways, like shopping at local businesses, helping your fellow travelers by offering sunscreen or a Band-Aid, and carrying the suitcase of that person with the aching knees. As a travel writer, I try to contribute by promoting locally-owned businesses and sharing the stories of people in remote places that rely on tourism. However you contribute, it's a more satisfying feeling than going somewhere and simply consuming resources and experiences.

> "I feel more confident talking to people and meeting people. When I was drinking, I felt like I had to hide. I knew I drank more than I should and I was really embarrassed about that. But the experiences I have when I travel now too are different. I'm not looking for bars or somewhere I can drink. I'm looking for places unique where I'm at."
> —Sober traveler Maddie, Denver, Colorado

WHY IS TRAVELING WORTHWHILE?

As Steve A., owner of Sober Vacations International puts it, "We didn't get sober just to have a small life." Travel helps us have bigger lives. Some of the benefits include:

- Learning more about ourselves and the world

- Meeting people in other places

- Strengthening relationships with family and friends who live in other cities or countries

- Growing your business by attending meetings and conferences in your field

- Developing more self-confidence

Sober trip idea: Oman

"A glistening, modern but exotic blend of desert, mountain and oceanside scenery, Oman surprises all who visit this small kingdom on the southeast corner of the Arabian peninsula. As in its UAE cousin Dubai, alcohol is available to foreigners in bars and restaurants (chiefly in hotels) but it is quite inconspicuous. Tours into the desert and mountains reveal ancient cultures—I had no idea there are as many different kinds of dates as I saw in a market in Nizwa—while modern glory is represented by the world's largest crystal chandelier in the central mosque in Muscat."

—Sober travel writer Eric Lucas

WHO DOESN'T DRINK?

I embarked on this project expecting to write a book for travelers in recovery from addiction. That's still my primary focus. But when I talked to people about my topic, I was surprised by how many identified as sober travelers who had never had a drinking problem. Consider my mind expanded!

While drinking statistics vary around the world, the majority of people in my home country of the U.S. drink. According to a 2019 National Survey on Drug Use and Health, 69.5 percent of American adults had drunk in the last year and 54.9 percent in the last month. That's a solid majority. But that still leaves 45 percent who haven't drunk in a month, and 30 percent who have foregone alcohol for at least a year. Who are these nondrinkers?

People who have quit a substance

Accurate stats are hard to come by when there's unfortunately still stigma around addiction. But one 2017 study[2] by the nonprofit Recovery Research Institute found that 9.1 percent, or 23 million, Americans have resolved a problem with alcohol or other drugs. Of these, 46 percent identified as being "in

---

2 "1 in 10 Americans report having resolved a significant substance use problem," www.recoveryanswers.org/research-post/1-in-10-americans-report-having-resolved-a-significant-substance-use-problem/

recovery." Fifty-one percent of the 23 million identified alcohol as their drug of choice, followed by cannabis, cocaine, meth, and opioids. Just over half sought help to overcome their addiction, while the other 46 percent quit on their own.

They also defined abstinence differently. Fifty-two percent said it meant not drinking or taking any other drugs, while 73 percent said abstinence meant only refraining from those substances they deemed personally problematic. Worldwide stats are even less reliable, with some countries being decidedly unsympathetic to the addict who needs help. One estimate puts the global number of people with alcohol use disorder at 300 million people, with alcohol being the cause of one in 20 deaths worldwide.

People taking a break from alcohol

Not everybody who develops alcohol dependency has a lifelong problem. Especially during the COVID-19 pandemic, many people noticed their consumption edging up. Their lives might still be perfectly manageable, but they worried about using alcohol as a crutch and possible health effects, and decided to take a break from drinking.

While the media often touts the benefits of drinking in moderation, people might be surprised by how the National Institute on Alcohol Abuse and Alcoholism (NIAAA) defines moderate drinking: up to one drink (12 ounces beer, five ounces wine or 1.5 ounces liquor) a day for women and two for men. After that, you get into disadvantage territory. Hence the rise of things like Dry January, Sober October and the sober–curious movement. Deciding to take a break for, say, 30 days gives you a chance to reevaluate your relationship with alcohol.

Gray-area drinkers

Gray-area drinkers fall somewhere between moderate and risky drinkers. They're not full-blown alcoholics but they sometimes question their drinking. They overlap with the sober curious crowd. They sometimes decide life would be better if they stopped drinking entirely, rather than keep trying to moderate, even though they may seem fully functional to others and be nowhere near rock bottom.

People who abstain for religious reasons

Some religions forbid the use of alcohol—not to say that everybody necessarily obeys, but many do. Islam forbids alcohol consumption, as does Jainism. Since Buddhism is about disciplining the mind, and alcohol produces the opposite effect, some practitioners vow to abstain. Some Christian sects also discourage alcohol use, while others forbid it, including Mormons, who mostly follow the rule. In the U.S., a 2019 Gallup poll found an inverse relationship between people who regularly attend religious services and those who drink.[3] The poll concluded that half of those Americans who attend weekly services are teetotalers. According to the poll, 18 percent of Mormons reported occasionally drinking. Southern Baptists, Seventh Day Adventists, and Nazarenes are also likely to abstain.

Baby-related reasons

Those who are pregnant, breastfeeding, or undergoing fertility treatments often abstain from alcohol use. This has been standard medical advice in the U.S. since 1981, when the surgeon general advised "women who are pregnant (or considering

---

3 "Religion and Drinking Alcohol in the U.S.," news.gallup.com/opinion/polling-matters/264713/religion-drinking-alcohol.aspx

pregnancy) not to drink alcoholic beverages and to be aware of the alcoholic content of foods and drugs."[4]

Other health motivations

Alcohol doesn't play well with some medications, including chemotherapy. Nor is it good for people with cardiovascular problems, high cholesterol, or high blood pressure. Anyone going through menopause can suddenly find what was once a pleasant glass or two of wine now triggers night sweats, hot flashes, and mood swings. Regular use can up your chances of developing various nasty cancers and, of course, liver disease. Nor is alcohol a friend to people who suffer from depression or other mental health issues.

> "I'm of Asian descent and experience an 'Asian flush' quickly, so I typically avoid alcohol to avoid the unwanted attention or appearance that I'm drunk when I've only had a few sips."—Mary Chong, food and travel writer and graphic designer, Toronto, Ontario

It just doesn't feel good

While the intoxicating effects of alcohol are popular in many places around the globe, some people value a clear mind and feeling fully in control. They may drink rarely or never. "What about those of us who simply don't drink because we don't like how it makes us feel in any dimension—physically, emotionally, or mentally?" asked Alaska-based travel and outdoors writer Lisa Maloney, who skips the drink 99 percent of the time. "I realize people don't talk about that reasoning much, but every so often I run across people who make the same choice for the same reason, and I think there might be more of us out here than folks tend to realize."

---

4 "Maternal Risk Factors for Fetal Alcohol Spectrum Disorders," www.ncbi.nlm.nih.gov/pmc/articles/PMC3860552

Not wanting to turn into Mom/Uncle Harry/ insert name of poorly behaved relative here

Alcohol use disorder is widely understood to be a family disease. Some folks have experienced so much drama and trauma in their families that they make the wise decision to not even start drinking in the first place. While they may or may not have inherited that addiction gene, is it really worth finding out?

Wanting to feel like yourself

As writer and podcaster Tasmin Waby wrote for Lonely Planet, "Perhaps a subset of the wellness movement, sober living is really about connection, rather than puritanism. By giving booze a raincheck on holidays, travelers are finding they feel more authentic, are more present, and start tapping into a sense of belonging that alcohol once temporarily provided."[5]

The weird idea that alcohol enhances everything

I tracked a leading Muslim travel writer down through the internet. He wished to remain anonymous but told me he likes to travel in Muslim-dominated countries because "the culture there will not be centered on drinking nor will there be an expectation that drinking will enhance an experience." Which is an interesting point, and has also been my experience in predominantly Muslim countries. In other parts of the world, it seems like many people think everything is better with alcohol, from boozy ice cream to the Disney Day Drinkers Club Facebook page, which currently has more than 101,000 followers!

---

5 "Is this the year to try sober travel?" www.lonelyplanet.com/articles/travel-trends-alcohol-free-tourism

People in positions of great responsibility

Some people might be called upon to perform at their best at any time—and they don't want to be caught buzzed when needed. "A friend told me that when she started working for a suicide hotline, she stopped drinking, because she never knew when she might be called upon to, well, save a life," Elizabeth Harryman Lasley told me. A fellow Society of American Travel Writers Sober Travel Affinity Group member, Elizabeth can always be counted on to help me brainstorm the ins and outs of nondrinking.

As you can see, there are many reasons people refrain from drinking. And they're all pretty darn personal. Whether you're in a 12-step group or recovering on your own, undergoing chemotherapy, a Baptist, or you simply don't like the way alcohol makes you feel, *The Sober Travel Handbook* aims to help you navigate travel situations as gracefully and stress-free as possible. You'll notice an emphasis in this book on people in recovery. This is because the stakes are highest for us. But I'm also striving to include plenty of useful info for other kinds of nondrinkers. Just skip through the parts about relapse and hopefully you will feel at home here, too.

Introducing Sober Cat

This book's first incarnation was as a zine, and zines need artwork. So Sober Cat was born. This nonbinary, hairless Sphynx lived a rough life in the alleys before getting sober. Growing up with neither fur nor an obvious gender identity (plus giant ears), Sober Cat always felt like the odd one out. They tried to drown their discomfort with alcohol and other drugs. During their misguided youth, they even got an unfortunate hate tattoo: Dogz Drool. Through the help of a sober mentor named Lucifer (full disclosure, Luce is my real-life cat), Sober Cat eventually got off the sauce. I hope that readers can relate to my illustrations of Sober Cat's feelings around alcohol and their responses to the many situations recovering people find themselves in. To me, a cat without fur seems as vulnerable and raw as an alcoholic without alcohol. If Sober Cat can thrive in recovery, so can we. I thought of this character after going on a sober trip to Costa Rica where two of the six participants coincidentally had Sphynx cats.

OUR HAIRLESS, NON-BINARY, SPHY[NX]
CAT RECOVERY HERO.

> *Sober trip idea: U.S. national parks*
> "They are kid-friendly and offer lots of activities for outdoor enthusiasts and nature lovers, which means that alcohol is not front and center. Plus, since you can pack your own food and beverages to take into the parks, you have more control over what you consume."
>
> —Anonymous sober traveler from Texas

RELAPSE TRIGGERS AND TRAVEL

efore we get to the fun stuff, we have to address the elephant in the room. For people in recovery, that elephant is the fear of relapse.

When the nondrinker travels—especially those who have quit because of a history of excess and regrettable behavior—we can find ourselves in unexpected situations that threaten our resolve. At home, many of us manage to maintain some sort of structure or control of our lives which helps keep us sober. On the road, anything can happen. That oft-quoted saying, "The price of liberty is eternal vigilance" are words the recovering person lives by. There's no letting your guard down with the certainty that what happens in Vegas stays in Vegas. Instead, what happens in Vegas is likely to follow you home and dismantle your reconstructed life.

The longer you're sober, the more you may tend to filter out messages about alcohol. At least for me, it's a survival skill that's developed naturally. Like a little kid who's trying not to hear the scary noises of a monster in the closet, I go through life with my fingers in my ears singing "la la la!" so I don't hear the world constantly whispering, "Drink, drink, drink." But it's always there, even on a lovely sober day, a kind of subterranean buzz of availability. And on a bad day, of seduction.

For those newly in recovery, this isn't an undercurrent but a tidal wave. Images of drinking bombard American culture and dominate many other countries as well. People clinking glasses in happy ads, inane wine mom and wine o'clock t-shirts, stupid jokes about it's five o'clock somewhere, happy hour specials. All these things can weaken resolve and help to trigger a relapse.

WHY WOULD TRAVEL TRIGGER RELAPSE?

Is travel especially likely to lead to relapse? And if so, why? I asked George Koob, director of the National Institute on Alcohol Abuse and Alcoholism, in a phone interview.

"Well, traveling is stressful," he said. "And stress is one of the major triggers for alcohol intake."

American culture associates alcohol with lowering stress and anxiety. And this works—up to a point. "It is effective for the hour or so that you're on the rising end of the blood alcohol curve," Dr. Koob said. "But shortly after that, the second drink doesn't do what the first drink does, and the third drink doesn't do what the second drink does." Obviously, people with alcohol use disorder don't stop at that first drink.

Opportunities to drink saturate the typical vacation, starting before you even get on a plane. "There's always bars in an airport," Dr. Koob said. "There's always things that are attractive that will remind you of the fun you had drinking. Then when you get on the plane, the person next to you may start drinking. Then when you get there, there's likely to be all kinds of triggers in the accommodations, including happy hours and free this and free that. It just goes on and on." His best advice for sober travelers? "Expect temptation. And I think if you expect temptation, then you're likely to be able to handle it a little better."

At my home airport of Portland, Oregon, Alaska Airlines has huge signs saying, "Now boarding rows chardonnay through pinot noir" as part of their "wine flies free" campaign, and a url where I can "uncork the details." To some, this is probably cute marketing. To alcoholics, this is part of the endless grind of going against the norm so that we may live with dignity for another day. Or live at all.

REWARD AND RELIEF

Some of the most enduring research and theory on relapse prevention was published in 1985 by Alan Marlatt and Judith Gordon. It's gone down in history as the Marlatt relapse prevention plan.

Marlatt and Gordon pondered the characteristics of situations that led to relapse. The main ones were:

- Negative emotional states, including anxiety, boredom, anger, depression, and frustration

- Situations involving other people, especially conflicts such as arguing with a partner or family member

- Social pressure, whether it's somebody saying something like, "Come on, you're more fun when you drink," or just being around others who are drinking

- Positive emotional states such as celebrations, or exposure to cues like passing a favorite bar and wanting to go inside and have just a drink—or 20[6]

If you read that closely, you noticed that two of the biggest relapse triggers are negative emotional states and positive emotional states. Uh, what's left? Are we doomed if we don't go around in a completely neutral mood all the time? No, but people in recovery need to tune into ourselves and notice when we are getting so high that we forget about our problem or so low that we just don't care.

Dr. Koob described these as the *reward craving* and the *relief craving*. You want to reward yourself and join in the celebration, or you want relief from your negative emotions. Both are likely to show up during travel. If you're on a beach surrounded by people holding umbrella drinks, you might want to join the fun. Or if your luggage got lost, you're sunburned, jetlagged, lonely, and wondering why you ever came on this stupid trip, you might feel like drowning those negative emotions. "If you're young

---

6 "Relapse Prevention, An Overview of Marlatt's Cognitive-Behavioral Model," https://pmc.ncbi.nlm.nih.gov/articles/PMC6760427

and a party animal, then it's probably more reward craving,"
said Dr. Koob. "But if you're older and have a history of severe
alcohol use disorder, it's probably more relief craving."

"Travel for both business and pleasure can be a risk factor
for substance abuse problems, for clinical deterioration in those
with chemical dependence, and for relapse when an addiction has
been in remission," some Johns Hopkins researchers concluded
in their article "Substance Abuse and International Travel."[7]

> "The main thing that contributes to any desire to drink
> or use is feeling shame, so I know that and check in with
> a friend or with myself about what I might be feeling
> ashamed about if any craving comes up. I think knowing
> one's triggers for drinking is really important."
> —Sugar, therapist, San Francisco Bay area

Celebratory relapse

Addicts can relapse because of depression, hopelessness, or
anxiety. But they may also relapse in celebratory circumstances,
telling themselves one drink can't hurt, you only live once, why
the hell not, fuck it, and other dangerous little phrases.

I had one of these celebratory relapses in the late 1980s. In
1987 I was a college sophomore living in New Orleans and
writing for my student newspaper. I heard that a local nonprofit
was giving scholarships to go to an enormous gay rights march
in Washington, D.C. The NAMES quilt, which commemorated
people who had died of AIDS, would be on display. There would
be a mass (not yet legal) wedding of gay and lesbian couples.
This was much earlier in the gay rights movement, and it
sounded like a phenomenal thing to cover.

---

7 "Substance Abuse and International Travel," www.tandfonline.com/doi/
abs/10.1300/J251v08n02_02

I got the scholarship and rode the train from New Orleans to Washington, D.C. At the time, I'd managed to not drink for about a week, which seemed like solid sobriety to me when I was 19. I'd booked a room in a guesthouse for a couple of nights in a residential neighborhood. While there, I met a guy who was also determined not to drink. Somehow we ended up playing cards together, then pretending we were sick and begging the lady who ran the guesthouse for some of her Nyquil. We drank the whole bottle.

The next day, I was running all over D.C. talking to excited queer people. At some point during the day, the celebratory spirit was just too much. Primed by the previous evening's Nyquil, I succumbed to the party atmosphere and started drinking. I have vague memories of a big gay disco and walking around by myself in the middle of the night in an unfamiliar D.C. neighborhood. The whole day I was carrying a bottle of fire engine red hair dye that I'd bought somewhere. Somehow I managed to find my way back to the guesthouse. In the wee hours of the morning, I decided to dye my hair. Then I went to bed.

The next morning, the proprietor was flipping out. "Who is hurt? Who is bleeding?" she cried, checking on all her guests. Apparently, I'd not been very neat in the bathroom, and it looked like the scene of a massacre. With my hair newly bright red, I denied all knowledge.

This is a tame and very lucky little relapse in that nobody died, nobody was hurt, and I managed to avoid being raped or assaulted as I wandered our nation's capital in a vulnerable state. But it illustrates how fragile, how silly we can be. Because other people were celebrating, I let my resolve dissolve. Because I teamed up with a similarly crooked mind, my fellow card player and I normalized for each other drinking a bottle of Nyquil. Because I was drunk and clumsy, I trashed someone's bathroom, not to mention scaring her that a guest was gravely injured.

Sober trip idea: New York Fashion Week Plan your next big city trip around the fashion event of the season, the Break Free Foundation's Recovery on the Runway designer showcase. Fashion designer and recovery advocate Alexandra Nyman founded the Break Free Foundation as a way to feature the work of designers in recovery, to raise awareness of behavioral health, substance use, and co-occurring disorders, and to serve as a fundraiser for the foundation's scholarship fund that sends individuals seeking to enter into recovery to a rehabilitation center at low to no cost to them. The show fittingly happens every September during National Recovery Month and every February right after Dry January. The runway show serves mocktails to all guests in attendance, and from their runway models to their designers to their performers, it is a celebration of recovery and recovering out loud. Alex is also the publisher of The Sober Curator website, which I write for. She is all in for the recovery community.

A negative state relapse trigger

Here's a story from Sugar, a therapist and my ex-boyfriend's sister, whom I've been close with for three decades. When Sugar visited Hawaii on her own, she faced more loneliness than she expected. It was spring break, and the islands were full of families and honeymooners. "It's usually really easy

for me to strike up a conversation with strangers," she said, "but there's no opening when people are preoccupied with their family or making out with their spouse." She felt like her singlehood was constantly shoved in her face. "It was really hard to be around so many families with young children that were driving me crazy, and then go to another spot and see nothing but canoodling couples. Even when I participated in day trip adventures, I would be the only solo traveler. It felt lonely and brought up a level of loneliness I do not often feel in other environments. The only thing I could imagine doing where I could strike up casual conversation with strangers was to sit at a bar, but I didn't want to drink so I didn't even feel like I had that option." Fortunately, she managed to avoid drinking. But it's important to remember that loneliness can be a huge trigger.

Family holiday gatherings and drinking

An anonymous source from Colorado told me about a near relapse while visiting a childhood friend's parents for Christmas. They all drank a lot "and thought giving me bitters would be okay. I did, too. I felt pretty uncomfortable with how much they all drank. I thought I'd feel better if I had just a little, but I knew that was a warning sign, so I called a sober person and the craving subsided. It might be important to know that this childhood friend I was visiting died drunk in her hot tub last February, and her mother died in a terrible drunk driving accident which almost also killed her father. These were all very successful people with good lives. I was 'the alcoholic' and felt like they felt sorry for me. But I'm still alive." Many of us have faced similarly uncomfortable feelings at family holiday parties.

Our problems travel with us

I came in contact with comic and writer Rebecca Rush when she was editing a website I wrote for. She shared many stories and tips with me that you'll see throughout this book. "As a stand-up comic who works the road, and a woman with a strong sense of wanderlust, travel is a big part of my life. For years I thought I loved the road more than anything—until I found myself in Chicago struggling to stay sober. In and out of recovery for nine years, that day it hit that I didn't love the road—I loved getting wasted with other comedians who didn't know I had a problem. If you've known me longer than six months, you've experienced one of my relapses. You KNOW I have a problem. That day in Chicago I realized I had no defenses against drinking on the road—I accepted victimhood to my alcoholism periodically anyway, and especially in that situation. I relapsed on that trip and started drinking after a show, and when a man on the street pulled a large bag of cocaine out of his pocket and winked at me, I wordlessly ditched the group and followed him, stepping out hours later into the cold light of morning with no idea where I was. Or who.

"In Thailand a few months later it happened again. Dehydrated and exhausted from travel, I rented a motorbike—I can't even ride a bicycle—and drove it directly into a copper wall. Limping down the street, I was sure I had earned a relapse. I bummed a smoke from an American retiree I had seen earlier that morning, and when I stood up, said, 'Well, I guess I better get good and drunk.' I looked up at the building. I was standing in front of the English-speaking AA clubhouse for the city of Chiang Mai, and the man was a member of the group. We went to the meeting instead. I stayed sober a few days thanks to that. But later, bored, not seeing the divine intervention for what it was, I started swiping on Tinder. Within hours, I was drinking beer with an expat from New Zealand."

Free booze!

Quite a few of us have experienced our cheap gene kicking in when the drinks are free. Several people I talked to while researching this book mentioned the temptation posed by all-inclusive resorts. Former gray-area drinker and sober coach Amanda Kuda of Austin, Texas, told me about a close call she had while visiting Mexico.

"I'm walking the resort and I'm thinking, 'Wow, I just screwed myself. I should have asked for a different package or something for nondrinkers. Because I paid for this. I paid for it.' And my sweet Midwestern frugal mind was spinning out of control." Fortunately, as a gray-area drinker, Amanda still had some standards. "And then I looked at the swim-up bar, and I saw what the all-inclusive alcohol was. And it was like pure lighter fluid. Even the top shelf. I was like, 'Ugh, that's not really all that great. I don't think that's so much of a deal.'"

Fellow travel writer Mary Chong, who I first met on a press trip to Jordan, told me she avoids all-inclusive resorts or package travel deals where alcohol is complimentary. "I don't wish to be around people who are drinking in excess to get their 'money's worth.'"

URGE SURFING

While people in recovery sometimes talk about the urge to drink being lifted, most of us will occasionally still feel the lure of alcohol. "Urge surfing" is one helpful visualization technique to manage cravings. Instead of identifying with the urge to drink as part of your innate desire, you visualize the urge as a wave, which rises and falls before crashing onto the beach. Instead of being drowned by the wave, imagine riding it on a surfboard. Or I like to imagine being out in the ocean where the swells are rising and wisely choosing which waves to ride and which to sit out. Either way, it helps if you can foster some detachment from your cravings, seeing them as things that rise and fall rather than emergencies that must be acted upon.

Sober trip idea: Provo, Utah
About 90 percent of this college town's 113,000 residents are Mormon, which means visitors will find a clean city full of polite people with little focus on alcohol. Instead of going out drinking on weekend nights, you'll see folks lined up at ice cream shops. Provo has a renowned dinosaur museum, as Utah is dinosaur central. I love the beautiful Rock Canyon Trail that starts at the edge of town. In summer, you can float the Provo River. Utah ranked number one in a study done by Lantana Recovery for best U.S. state for sober travel, based on availability of sober-friendly activities, low alcohol consumption rates, alcohol laws, public safety, and other factors.[8]

---

8 "The Sober Travel Destinations Index," lantanarecovery.com/sober-travel-destinations-index

ARE YOU SOBER ENOUGH TO TRAVEL?

When I got sober in the late 1980s, it was more unusual for somebody my age to be in recovery than it is now. I don't know how many times people said things like, "But when your friends get married, surely you'll be able to toast them with one glass of champagne." Uh, did you never notice that water in a glass makes the same clinking sound?

Unfortunately, this attitude is still prevalent and produces a lot of anxiety in nondrinkers. Things might be fine when we stay in our own controlled environment, doing whatever we do to take care of our sobriety, whether that be therapy, recovery meetings, church, yoga, time in nature, a network of sober friends, or what have you. But what would it be like to go somewhere where you'll be the only person not boozing it up? The person who toasts a wedding with water, who skips out of the late-night parties, who just says no on the cruise line when a server comes around with a tray of umbrella drinks? Just thinking about these awkward moments is enough to keep many nondrinkers from venturing outside their comfort zone.

For people whose mental, spiritual, and physical health has crumbled from substance abuse, sobriety is the most precious thing. It's much more important than traveling. If you're new to sobriety or going through a fragile phase, staying home where you feel safe is a good call. But once you've built up your sobriety muscles, you might start to wonder if you're sober enough to travel.

"When it comes to individual recovery, everyone is different," said Sarah O'Brien, addiction specialist with Ark Behavioral Health in Massachusetts. "I don't believe there's a certain amount of time or a length of sobriety or the amount of programs or treatments that you've completed that deems you ready to do these things."

Instead, Sarah urges people to note how they feel on a day-to-day basis. "What does that look like in terms of thoughts or cravings that could potentially be heightened if you travel alone or you travel to a place where maybe drinking is a little more broadcast?" She recommends people talk to someone with whom they've grown a relationship in sobriety, whether that be a sponsor in a recovery-based program, a fellow alcoholic or addict they've bonded with, a therapist, or a family member. "Express your hesitations or fears and see what others think." Maybe they'll tell you this isn't the right time. Maybe they'll tell you it's a really great time, they think you're ready. It's up to you what you do with the suggestions. But sometimes other people see us more clearly than we see ourselves, especially in early sobriety.

I talked to David Smith of Dallas when he was eight months sober, and just starting to travel without alcohol. He recommended recovering people stay sober for at least five or six months before venturing out. "Even if it's a completely sober trip, there's still lots of opportunities for drinking in the airport bars," he said. You're still likely to transit through situations where people are drinking, even if you're traveling with fellow nondrinkers.

Also, he warned, "If they're traveling by themselves, people have the opportunity to possibly drink without anybody else knowing. But ultimately you're doing this for yourself. So you know if you've done something wrong." He suggests experimenting in your home environment to see if you're okay having dinner with a friend who drinks, or going to somebody's house where people are drinking. "So my recommendation is: challenge yourself in a healthy manner at home. And if you find that you're strong enough in home situations, then you may be ready to travel."

Sober trip idea: Ahmedabad, India

"One place that is an ultimate destination for a sober trip is Ahmedabad, India," said Martin Retch, co-founder of Hi-van, a van life website. "First of all, it is a completely dry city and there is no chance of an alcoholic encounter anywhere. The policy is implemented very seriously. There is also a de-addiction center which is properly managed and run by a well-established organization.

The best part about the place is its historical preservation and importance, he said. "Kankaria Lake, Sabarmati Ashram, Sidi Sayyed Mosque, Bhadra Fort, and the Adalaj Stepwell are major tourist attractions which have played a vital role in the country's independence. The architecture and sculptures of the buildings are exquisite and antique. Also, the place has its unique cultural cuisine. A person who is interested in sober travel will definitely love the aura and vibe of the place."

DISULFIRAM TO THE RESCUE?

Disulfiram is a drug that makes you sick if combined with alcohol. This effect was discovered accidentally in the 1930s when rubber industry workers using tetraethylthiuram disulfide became ill if they consumed alcohol. And so a treatment for alcoholism was born.

Denmark and Sweden first approved disulfiram in 1949. In 1951, the FDA approved it for Americans.[9] Disulfiram, often sold under the name Antabuse, has been used around the world ever since to strongly discourage people from drinking. It seems a little extreme, but one sober traveler suggested taking it temporarily for those worried about succumbing to temptation while on a trip.

I took Antabuse for a few months when I first quit drinking. It was both extremely helpful and humiliating, especially the part about being legally required to carry a card explaining I was on disulfiram therapy and if I was found unconscious not to administer alcohol. (I was living in New Orleans, so perhaps this seemed like a reasonable medical intervention.) But if you look at it objectively, this might be a useful tool for somebody who really wants to go on a trip and is afraid of relapsing.

In 2023, there was a worldwide disulfiram shortage and some manufacturers had ceased production, so by the time you read this that may or may not be resolved.

SELF-EFFICACY

Addiction researchers have long tried to gauge a recovering person's stability and their likelihood to relapse. One of the ways they do this is by that favorite tool of the alcoholism industry: the questionnaire.

9 "From Disulfiram to Antabuse: The Invention of a Drug," acshist.scs.illinois.edu/bulletin_open_access/v33-2/v33-2%20p82-88.pdf

Most people who have ever questioned their relationship with alcohol have probably taken an "Are you an alcoholic?" quiz online, or, if you've been sober (or trying to get sober) as long as me, on a paper pamphlet. But when assessing how successful the sober person could potentially be, therapists may encourage them to take a self-efficacy quiz.

So, what is self-efficacy? It's not a term that came up for most of us while sitting on a barstool. According to the American Psychological Association, "Self-efficacy refers to an individual's belief in his or her capacity to execute behaviors necessary to produce specific performance attainments. Self-efficacy reflects confidence in the ability to exert control over one's own motivation, behavior, and social environment. These cognitive self-evaluations influence all manner of human experience, including the goals for which people strive, the amount of energy expended toward goal achievement, and likelihood of attaining particular levels of behavioral performance. Unlike traditional psychological constructs, self-efficacy beliefs are hypothesized to vary depending on the domain of functioning and circumstances surrounding the occurrence of behavior."[10]

Self-efficacy questionnaires for people recovering from substance abuse disorder ask you to imagine yourself in a situation and then estimate whether you'd resist picking up a drink or other drug. This theory recognizes that our resolve will vary depending on the circumstances. Typical examples from an alcohol questionnaire include very general emotion-related situations, such as: "If I were angry at the way things had turned out," and more specific cases, such as: "When I'm on vacation and want to relax." The quiz taker rates her confidence from 0 to 100 whether she'd be able to resist the substance.

10 "Teaching Tip Sheet, Self-Efficacy," www.apa.org/pi/aids/resources/education/self-efficacy

Self-efficacy is a skill we can learn and strengthen. Spending time building our self-efficacy muscles pays off. While the high-risk situation—whether partying or despair—is the immediate relapse trigger, it's how we respond to the situation that dictates whether or not we relapse. That's why we need to develop our coping skills and ponder temptations before they happen. These skills include behavioral strategies like having an exit plan to leave an event if necessary, and cognitive strategies like positive self-talk.

People who have quit alcohol must participate in the world without anesthesia. It's like a turtle having to do without a shell. But when people practice dancing, socializing, dating, working, driving, traveling, and doing everything else sober, our self-efficacy grows. Developing confidence in our ability to meet the world sober is crucial if we want to have a bigger life than going to work, going to therapy and/or a 12-step meeting, and going home.

Sober trip idea: Join a group hiking trip to a national park

"Last year I went hiking in Glacier Park in Montana, and what a beautiful place!!" said Sarah Michaud, a clinical psychologist, sober traveler, and fellow contributor to *The Sober Curator.* "I went with an organization called Off the Beaten Path and it was such a great time. First time I ever went with a group of people I didn't know. If you like to hike, what an incredibly beautiful and magical place. Alcohol was definitely not the focus, even though we all got together for dinners at night. You were so exhausted by the end of the day that all you wanted was to have a great meal and pass out in bed. I had always wanted to go to Montana—and they call it Big Sky for a reason."

PREPARING FOR DEPARTURE

MANAGING EXPECTATIONS

People can put a lot of pressure on their vacations. Travel brochures and social media emphasize bucket list travel, trips of a lifetime, etcetera. Planning can be a lot of the fun of travel—searching for the hotel closest to the beach with the biggest pool, or the remote primitive cabin with the most pristine lake, or golfing at one of the world's top courses, or whatever floats your boat.

But having too many expectations of a place can ruin your vacation. The pool might be closed. Your hotel room is smaller than it looked on the website. An Uber driver yells at you for waiting at the wrong entrance of the hotel. There could be mosquitos. All these things can disproportionately disappoint alcoholics and other dramatic people.

Then there's what we expect of ourselves in these beautiful places. We look at these gorgeous people in the website photos and think we'll feel as at ease in our swimsuits as they do. Sleeping in a tent after a 10-mile hike will feel super. The resort will be full of friendly people and maybe we'll even find romance. And these things might well be true. Then again, we might pull a hamstring or agonize over stretch marks or feel shy or grumpy and not want to talk to anybody.

Instead of expecting so much of places and of myself, I try to be a bit more open-minded about what I'll find and how I'll feel. For example, when I get up super early and fly to the East Coast without sleeping much, I'll probably arrive at a hotel in a bad mood, wishing I hadn't left home and pretty much hating everybody. That means I should eat something nutritious and go to bed early. The next day I'll probably feel better and remember why I came. This may sound like advice for a toddler, but it's a real pattern where I'm exhausted and bewildered about what I'm doing here in this hotel room far from home, and maybe I don't deserve to travel because I don't appreciate it, and maybe I'm a worthless human being, etcetera. When really, I just need to sleep. If you've ever been in a therapy group where they talked about HALT—don't let yourself get too *hungry, angry, lonely* or *tired*—you'll recognize the problem here is T.

Vacation is a challenging time for gray-area drinkers—those who have some concern about their drinking patterns but don't identify as alcoholics. Amanda Kuda is a sober coach who works with gray-area drinkers, and is the author of *Unbottled Potential: Break Up with Alcohol and Break Through to Your Best Life.* I found her online when researching this book and emailed her asking for an interview. She opened up the world of gray-area drinking to me, and had an astonishing number of helpful tips.

"I think a lot of people, especially if they're kind of dabbling in the sober-curious area or they're abstaining or taking a break, there always is this question about well, what about vacation?" said Amanda. How can they visit Italy and not drink wine? What's Mexico without tequila or Russia without vodka? "Because socially we're told that vacation is a time to cut loose and overindulge. And most people have this big fear about it. So they set it aside as kind of a negotiable period."

Instead, she said, the novel thing to do would be to have a clear-headed sober vacation, experiencing a new place sans hangovers. "Everyone who I work with who has ever chosen to take that path has always been so astounded. I've heard them have moments of disappointment. But in the end, the level of pride and just peace and excitement that they feel completely outweighs those moments of anxiety when they really wanted to give in to that social pull to have a drink."

This wonderful new world of sober travel might not come naturally. That's why we need to prepare ourselves before traveling: by packing useful stuff, coming up with a few plans B and C, and managing our expectations.

Sometimes nondrinkers just have to accept we'll be a little different from the norm. Sally, creator of the Passports and Plates website, is a Muslim who abstains for religious reasons. When she was traveling solo in her twenties, she noticed the divide between drinkers and nondrinkers. "I often stayed in hostels and met people there or via activities like walking tours, etc. I was a budget traveler, as were/are most people staying in hostels, and found that most people were interested in the nightlife aspect of traveling. I didn't really feel comfortable in that scene so sometimes I'd miss out on hangouts because of that." However, this discomfort didn't stop her from traveling and becoming a professional travel writer.

Sober traveler Rebecca Rush recommends Bali for so many reasons: hospitality, ubiquitous yoga, low prices, ease of getting around, and the fact that it's perfect for sober travelers, including solo women. "Bali is a Hindu island," Rebecca wrote in an article for Fodor's. "Hinduism is a fascinating religion, and intoxicants are discouraged. A land without drunk men or drug dealers is exponentially safer, especially for solo ladies. In my personal experience, I never saw Balinese people drink alcohol. No one ever offered or tried to sell me drugs. I never experienced catcalling as I walked down the street after dark."[11]

THINGS TO PACK TO KEEP YOU SOBER

Packing is preparing. Try to anticipate what items will help you live your best sober life on the road.

Reading material

A book can be a friend when you travel. If you're alone, it provides company. If you want to retreat from others, it offers escape. Addiction specialist Sarah O'Brien says she always brings prayer or meditation books that she reads on a daily basis. "I try to keep things very similar at home as I do on the road, whether I'm on a beach or traveling for work," she said. Bring your favorite spiritual or recovery literature and you won't miss a day. Sober traveler Eric Lucas packs his meditation

---

11 "10 Reaons Why Bali is Perfect for Solo Women Travelers," www.fodors.com/world/asia/indonesia/places/bali/experiences/news/photos/10-reasons-why-bali-is-perfect-for-solo-women-travelers

book. He also suggests checking out the Hazelden/Betty Ford website to find daily thoughts for people in recovery.

Personally, I prefer a page-turning novel. I have a lifelong need to disappear for periods of time—hence my former frequent trips into alcoholic oblivion—which I now do by mind melding into the lives of characters in gripping books. I always regret when I go on trips and only pack books that I think are in some way "good for me." This usually results in expensive book purchases at airports.

"I definitely stay in touch with my alcoholism. I don't let that become something invisible," says sober traveler Sarah. "The most important one for me is staying in touch by reading. Like, I think some people do a lot more socializing or calling people on the phone. For me it's staying aware of who I am, what I'm doing, why I'm doing it, and that I'm an alcoholic wherever I am is really important. And the way for me to do that is to keep in touch with the daily readings. I always take my readings with me."

Something fun

Do you like crosswords, jumbles, or sudoku? Favorite games on your phone? Sometimes you need to turn off parts of your mind as a self-soothing behavior. For me, it's Bejeweled, especially on an all-night flight where my concentration is too worn down to follow the plot in my book.

Something constructive

If you're on the driven side, like me, you might actually enjoy bringing work on vacation—as long as it's work you really want to do. I'm always daydreaming about having more time to devote to writing projects. When I fantasize about the perfect vacation,

I picture half of it spent seeing the place and half in a beautiful hotel room, working on my laptop. I spent most of September 2018 traveling alone in Sri Lanka, splitting my time between sightseeing and writing. It was heaven. Maybe your version of "constructive" is bringing your sketchbook, practicing a foreign language on Duolingo, or brainstorming fundraising ideas for your nonprofit board. Others might shake their heads at you, saying you need to leave all that at home. The trick is to get to know yourself as a traveler.

Something comforting

Once I was on a solo trip to Cambodia and was feeling a bit disjointed and anxious. I found a copy of a Sue Grafton alphabet novel and spent hours in the familiar world of detective Kinsey Millhone. It was extremely calming to be with a fictional character I knew well, who could be relied upon to solve the crime.

As Renee Deveney, former content manager for the behavioral healthcare company Advanced Recovery Systems, puts it, "Comfort is essential to recovery. Make sure you have coping essentials on hand to counter potential setbacks. If a certain pillow, for example, makes you feel a heightened sense of security, bring that pillow with you on the road. Being in recovery doesn't mean you should nitpick or be overly concerned with packing light. If you think you might need it, bring it."

Food

Packing some food is super important. Ever get hangry? Not a good way to have a fun vacation or make a first impression on a business trip. Bring something nutritious and easy to pack, like protein bars. If you have a special diet, bringing food is even more crucial. I'm vegan and feel much more prepared when I devote a section of my suitcase to nuts, granola bars, and even home-baked banana bread. Just be sure to put all your food together and maybe even take it out before it goes through the airport x-ray machine. This is advice I always forget to take—my bag has been searched more times than I care to admit for mysterious vegan food items. Even after writing this, during the editing process of this book, I was detained for tofu.

Chargers and adapters

We live in the age of chargers, and it's very frustrating when you forget them. For most of us, the phone charger is number one, since we use phones for just about everything these days. Sober travelers need a charger for keeping in touch with our support networks. So double- and triple-check that you packed it. Then inventory any other chargers you need, like for laptops, fitness trackers, digital cameras, e-readers, and any other favorite electronic devices. Plus: adapters if you're going abroad or staying with relatives who only have two-prong outlets that won't charge your laptop.

Stamps

Want to make somebody's day? Send them an honest-to-God postcard. If you're traveling domestically, bring some stamps along to make this easier. Writing postcards in your hotel room can also be a good exercise in gratitude, as you jot down a few highlights of your day and perhaps feel lucky you can travel.

Journal

Journaling is always a good idea for nondrinkers, and anybody else with feelings. You can pour it all out to Dear Diary. Especially if it's two in the morning and you feel a little uneasy, but not bad enough to wake up your support system. Journals are also a great way to track moods and patterns, and to jog your memory later about all the beautiful, quirky, moving, and/or mystifying things you encounter on your travels.

"I think the best tool to have with you is some sort of journal," said sober coach Amanda Kuda. "Vacation creates this space of free thought. And when we get into our free thoughts and we have a little extra space to think, sometimes uncomfortable thoughts are what come up. And that's typically why we drink."

Lacking the capacity or the desire to deal with these uncomfortable thoughts, Amanda says, we tend to quiet our minds the quickest way we know how—with a drink. "But the problem with that is if our mind wants something to be heard, it's going to keep bringing it up, over and over and over." Instead, she recommends taking a short break. Walk somewhere quiet. "Get the thoughts out of your head so that they stop taking up space. Pack a place for you to put your thoughts."

Sober podcasts

Download some sober podcasts ahead of time, and you'll be able to find inspiration anywhere you go. "Before a flight, I walk around the airport, listening to the wise words of my people," Sober Outside founder Brooke Morton wrote on her website. "This keeps me happy and filled up, as opposed to eyeing the bar."

> "Everything's on my phone now. So my recovery literature is on my phone. I always have that with me and it's downloaded, so I have that accessible all the time. I just find it a little easier than bringing a book with me."
> —Nicole Kale, sober traveler from Charleston, SC

WHAT DO YOU DO ON VACATION IF YOU DON'T DRINK?

Who needs to find a nondrinking idea for their next trip?

• People newly in recovery, who may have lost track of interests they once had before substance cravings took over their lives

• People whose vacations have usually centered on wine tasting, night clubbing, or other alcohol-intense environments, but who are now taking a break from drinking

• People experiencing health changes that don't mix well with alcohol

• Nondrinkers who yearn for meaningful travel—especially if they tend to get roped into vacations with heavy-drinking friends or family

• People interested in health, wellness, and spiritual growth

> "An individual with an addictive behavior problem has a life out of balance to the extent that he or she is relying almost exclusively on a particular substance or activity for gratification and coping with stress. One thing that has struck me in working with clients with these behavior problems over the years is that they rarely have time to give themselves anything very positive in their lives outside of the addiction… In giving up an addictive behavior, an individual has to look at other ways of coping or self-reward as alternatives to getting intoxicated."
> —G. Alan Marlatt, psychologist and addiction researcher

So, what are your interests? For some of us, it's easy to immediately think of 20 things that we like to do. Other people may have trouble coming up with three. Take a few minutes and brainstorm a list. It doesn't matter how eclectic your list is, and whether your friends would think your interests are cool or dumb. Here's an example of the first 20-some things I can think of:

- Old cemeteries

- Vegan ice cream

- Public gardens

- Places to run

- Kayaking or paddle boarding

- Swimming

- Nature and wildlife

- Yoga

- Unusual museums

- History museums

- Local recovery meetings

- Old Catholic churches

- Ghost walks and haunted places

- Hot springs

- Scuba diving

- Spas

- Cat cafés

- Vegan restaurants

- Bookstores

- Garage sales

- Hiking trails

- Historic prisons and internment camps

With a list like this, you can see why I love solo travel! Not every travel partner wants to get up early to go kayaking, then tour an old internment camp, share some purrs at a cat café, and top it off with vegan ice cream.

If you look at my list, some themes quickly pop out. It looks like I'd like to visit a historical place with old cemeteries and churches, and perhaps a ghost walk. Or I could plan an outdoorsy vacation with hiking, kayaking and wildlife viewing. Or a trip to somewhere with a big vegan restaurant scene, like Santa Cruz, California, Brighton, England, or Berlin.

Whatever your interests, own them and think about incorporating them into your next vacation. Do you like birds? Maybe you should check out the Texas Gulf Coast or Costa Rica. A spa vacation? There's a massage with your name on it, somewhere beautiful and relaxing. Do you enjoy reading on the beach? The world is full of beaches. One of the keys to a successful sober vacation is thinking about what *you* want to do—whether or not your usual travel companions share the exact same interests.

In his book *Positive Addiction*, William Glasser defines a negative addiction as something that initially feels good but harms you in the long run. A positive addiction, on the other hand, doesn't immediately feel good but pays off over time. For example, running or meditating might not seem fun the first few times, and learning the guitar is painful on your fingertips, but these all provide big payoffs once you get hooked on them.

George Koob stresses the importance of thinking ahead about how you'll spend your time on vacation. "Make a plan, from getting your boarding pass to returning home," he said. "What are you going to do that's alcohol-free, if that's the way you want to be? Find activities that do not involve, to put it bluntly, boozing up." He suggests hiking, sailing, snorkeling, and yoga as examples of activities that are conducive to sober living. After all, who wants to carry a fifth of vodka when they're climbing a giant hill?

For those in early-ish recovery, there could be some trial and error about what you like to do now. Make a dream list of things you think you *might* like. Be open-minded. Try them once or twice before you decide whether or not they're keepers in your new sober life. As Darci Murray, owner of the tour operator Hooked Alcohol-Free Travel put it, "So we used to spend six hours a day drinking. Now we have all this free time. Who are we? We're kind of like this blank slate. Let's go to a different country, experience different healthy habits and see who we are. What resonates with us? Who the heck knows? Maybe we like birdwatching. You have such a different appreciation for things when you're sober, right?" While traveling in Tanzania, Darci found herself unexpectedly fascinated by birds. "I would have laughed at birdwatchers before," she said. Darci started out ashamed that she was in recovery, and parked her car blocks from meetings. Now she owns an alcohol-free travel biz and is a role model for those aspiring to try new things and live their sober lives to the fullest.

Sober traveler David Smith suggests that people venture beyond their usual vacation spots. "Be willing to go outside of your comfort zone and not just book a place because you've been there several times. Because the person that used to travel there is not the same person that's traveling now."

Of course, some places offer way more than a party scene and are worth revisiting. Maybe you've seen half the bars in Europe but slept through opportunities to visit art museums. "I like using the expression 'travel do-over,'" Darci said. "A lot of us have traveled in the past. What have we experienced? I've been to Italy before, but I've seen Italy at night time. I was too hungover during the day, and I wasn't really interested."

CONTRARY TO FELINE STEREOTYPES, SOBER CAT AND LUCIFER LOVED SNORKELING IN THE GALAPAGOS.

 assuming you like rides and don't mind crowds. One of my favorites is Silver Dollar City, an 1880s-style theme park near Branson, Missouri. It all started with the massive Marvel Cave, which you can tour. About 40 rides and attractions grew up around it, plus a crafts colony with 100 artisans demonstrating old-timey crafts. I love the fascinatingly strange and site-specific rides like the Fire-in-the-Hole indoor roller coaster, based on a local chapter of vigilante history. Or the Flooded Mine, where you're supposed to be helping the warden because the county prison mine is flooding and prisoners are trying to escape. Pure Ozark fun!

CAN TRAVEL BE AN ADDICTION?

id you know that there's a word for travel addiction? According to the American Psychological Association, dromomania is "an abnormal drive or desire to travel that involves spending beyond one's means and sacrificing job, partner, or security in the lust for new experiences. People with dromomania not only feel more alive when traveling but also start planning their next trip as soon as they arrive home. Fantasies about travel occupy many of their waking thoughts and some of their dreams. The condition was formerly referred to as vagabond neurosis."

I have definitely had bouts of this. Sometimes while traveling, I'm homesick and idealize getting home to a more regular routine. But I'm not back two days before I'm scrolling airline sites thinking about somewhere else to go. Or it's my work as a travel writer that necessitates thinking of one place while I'm in another, either because I have a deadline or because I need to make arrangements for the next assignment. Like the time I was on a ship in the Arctic trying to use the erratic internet to book flights to the Grand Canyon. And while getting hyped up about travel is way better for you than the demoralizing and often deadly results of alcohol and other drugs, it's still something to ponder if you want to be present in your travel and life.

Why is travel so addictive? It's two-fold: You escape the everyday problems of home, while getting constant dopamine hits from new experiences. "Though travel can be a wonderful avenue to presence, it can also do just the opposite," Breanne Kiefner, founder of Root Adventures, pointed out. Root is a group adventure travel company that is based in mindfulness and bills itself as "travel that makes you pay attention." I met Breanne at an adventure travel event in Asheville, North Carolina, and immediately connected over the way we want to see the world.

"Intentional travel fueled by curiosity can certainly help us connect to the present and help us grow," she said in a later phone conversation. "However, travel can be a distraction from 'real life,' allowing people to escape the trials and tribulations of daily life. And just like any numbing agent, your problems and frustrations will be waiting for you when you return or come down from the high."

Travel journalist Maggie Parker described travel addiction well in a 2014 article for Yahoo Travel: "I enter into hibernation, where I don't leave my house for a week, claiming I need to 'recover' from the exhausting trip I just had, but in truth it's a lot more than that. I feel like I'm in a black hole and I cry a lot. About what, I couldn't tell you—it's different after each trip. I miss the people I met, the culture I learned about, the things I did. I need to get back on a plane again soon, even though I wouldn't be going back to the place I'm missing or to see the same people. I just need to feel that high again."[12]

While most people don't experience this degree of travel addiction—and most people who don't work in travel can't

------

12 "Depressed After a Trip? Travel Addiction is a Real Thing," www.yahoo.com/news/its-true-you-can-be-dangerously-addicted-to-traveling-94436835972.html

afford to take that many trips—it's something to keep in mind. Instead of going for more and bigger travel experiences, slow down and savor the ones you have. "When I think of the powerful effects of travel, I like to remember that travel doesn't have to mean a *big* trip; it means staying truly present and embracing all the beauty life has to offer," said Breanne. "Healing comes from connection and connection can only be established with presence. Just as 'the unexamined life is not worth living,' I believe that the unexamined trip is not worth taking."

Sober trip idea: A running holiday

Calgary-based sober hospitality professional Makina Labrecque took up running after she quit drinking. "My first trip that I took as a nondrinker was to Monterey Bay in California. That was for my first big half marathon that I ever ran. It was a really life-changing experience." She loved being in such a beautiful, beachy place, where she ran and went whale watching.

"I think the health and wellness culture is really top-of-mind in California," Labrecque said. She loves that running in different places includes built-in sightseeing. "In adult life I think it's important to find a hobby you can travel for. If it's fitness related, better. I think that's the best way to explore new places."

YOUR SUPPORT SYSTEM ON THE ROAD

SUPPORT STARTS AT HOME

Many travelers in early recovery worry about leaving their support system. It's understandable. You might have put a lot of time, tears, and hard work into building trusting relationships with new friends—perhaps other people in recovery—and rebuilding relationships with old friends and family. You might have slogged through hours and hours of therapy getting yourself on track, and it can be nerve-wracking to put several time zones between you and your therapist. Or maybe you rely on other good things you've added to your life, like church friends or your book club or running group or gym buddies that keep you focused on the positive.

Fortunately, thanks to modern technology, you don't need to leave them at home. Load up your contact list with supportive friends. If you're in recovery, other people who have grappled with addiction will best understand what you're going through. But don't limit yourself—everybody has had troubles, and many will be sympathetic even if their problems don't identically mirror yours. Members of your support system don't even have to be nondrinkers, but they must be people that support your decision not to drink. And those who can take it or leave it are a better bet than people for whom alcohol seems to be a habit.

PREPPING YOUR SUPPORT SYSTEM

Once you've thought it over, make a list of who's on your side that you think will support you during your trip. These should be people you feel comfortable with, who know about your decision not to drink. Then prep them. Here's a sample script: "Hey, I'm going on a trip. Do you mind if I check in with you every couple of days, just to help keep me safe and sober?" If they say yes, confirm how they'd like you to contact them (if you don't already know). Depending on the person's relationship with technology, they might want you to text, email, call, or use WhatsApp. Get their mailing address, too, so you can surprise them with an old-fashioned postcard. They might not have received one for years.

When you're traveling, you may or may not have any close calls with alcohol. If you do, you'll have your support system on standby. But if you don't, you should still commit to staying in touch. You don't even have to mention sobriety, if you don't need to. "Stay connected with your support group, even if it's digitally. 'I just got here on (whatever island you landed on) in the Caribbean, everything's great.' If you have a sponsor, send it to your sponsor. Could be anybody. Could be Mom. Could be your brother, your sister, whatever," said George Koob.

FINDING AN ALLY ON YOUR TRIP

When I find myself in a group travel situation with people I don't know, I sometimes look for an ally. Ideally this would be a fellow nondrinker. You might notice somebody else is turning down the wine. Maybe someone will question them, and they'll say something that hints at recovery, or even mention it outright. Then you can discreetly tell them at some point—preferably not in front of other people—that you are also in recovery.

I was once with a group of writers touring a haunted hotel. One especially aggressive ghost was said to haunt a particular room. It happened to be a room that one of our writers was staying in, and he invited us all to crowd inside and check it out. I didn't see any ghosts, but I did notice a bunch of AA chips on his keychain on a table. AA or NA chips bode well when looking for a nondrinking ally.

Other potential nondrinking allies include folks who don't drink for religious reasons, people who are pregnant or on medications that interact with alcohol, and health nuts. If you don't have all that much in common besides not drinking, no problem. Try to pal around with them anyway, if they don't mind, and you'll expand your horizons by getting to know somebody different from you. It's especially nice at longer formal meals to sit beside someone who also turns their wineglass upside down.

Normal, moderate drinkers also can be good allies. People who work in healthcare might be clued in to addiction problems. Many people have a friend or family member who's gone through treatment and might be supportive. Also, anybody who describes herself as an empath will probably want to support you. I've had success with nurturing women who tend toward the busybody and gossipy. If you confide in them that you had a drinking problem, they'll watch you like a hawk and spread the word for others to keep their bottles away from you. Sure, it's a little embarrassing. But sobriety is way more important than looking cool.

CHECK OUT LOCAL RECOVERY MEETINGS

When we travel to another place with an unfamiliar culture, it's easy to think maybe we could bend the rules a little and be someone else. Attending any of these groups helps you remember who you are and why you decided to stop drinking.

Many people recovering from addictions find support in peer-based groups. There's no reason you can't do this while traveling. The beauty of attending recovery meetings when traveling is you meet other people with at least one thing in common, who are generally friendly and welcoming. You might even find somebody you'll want to get coffee or lunch with during your stay.

"Before I go to a place I always look up recovery meetings and kind of have a plan for that," sober traveler Nicole Kale of Charleston, South Carolina told me during a group sober trip to Costa Rica. "I can't say I always go, especially if I'm only in a place for a day or two, but I always know where they are. So if shit hits the fan, I know exactly where I need to go." Going to recovery meetings in other countries has enriched her travel experience as well as helping her stay sober.

Finding a meeting in an unfamiliar locale can be a memorable part of your trip. You'll likely find yourself wandering around the outside of some old building, looking for the alcoholics' entrance. You might even have to ask somebody where the meeting is, which can be a humbling experience. Most self-help groups are volunteer-run, so there's always the chance of delays or canceled meetings that haven't been updated in the schedule. Bring your patience and sense of humor and adventure with you.

12-step groups

You'll find meetings of Alcoholics Anonymous and Narcotics Anonymous around the world. Right now, there are more than 123,000 AA groups meeting in 180 countries, with an estimated two million members. If you embrace the 12-step approach, great. But even if you're ambivalent about AA, it can save your butt when you're far from home and on shaky ground, so keep the hotline number handy.

Secular recovery organizations

Like any organization that has grown huge and successful, people find plenty to criticize about the 12-step programs. Especially their reliance on a higher power. So various secular groups have sprung up, such as SMART Recovery, Women for Sobriety, Secular Organizations for Sobriety, and LifeRing. These groups may be small compared to AA, but they have a track record. Women for Sobriety was founded in 1975 and Secular Organizations for Sobriety has been around since 1985. SMART Recovery has about 3200 meetings worldwide, with 2000 in the U.S. Each of these programs is much smaller than the anonymous giants, but worth checking out if you crave a secular approach.

Faith-based recovery groups

Then again, maybe "higher power" is too vague for you. Celebrate Recovery is the biggest Christian self-help organization, with about 17,000 groups worldwide. It tackles other behavioral problems in addition to addiction. If you're in the right place, you can also find meetings of smaller faith-based peer recovery groups, such as Jewish Alcoholics, Chemically Dependent Persons and Significant Others (JACS), and Millati Islami, which is a 12 step-based group for Muslims.

Online meetings

There's also help virtually, especially since the pandemic forced organizations to rely on technology more than they had in the past. This means as long as you have an internet connection, you can access a 12-step, secular, or faith-based meeting on your phone, laptop or iPad. For some of us, this isn't as satisfying as in-person contact. But others prefer the convenience of virtual meetings. One of my friends only attends online meetings. I don't think she's ever met her sponsor in person, which she is totally okay with. Even if you're the kind of people-person who wants to get together for real, a virtual meeting can be a whole lot better than nothing.

> "If you are a person who goes to some sort of recovery meetings, get to those. And get to them within 24 hours of arrival. There's something magic in that first 24 of a trip that sets the tone. Set your compass away from a drink or drug by connecting to the sangha that keeps you sober, whatever it may be. If you're going to be away for longer than a week, temporary sponsors can be an amazing thing. Fellowshipping will give you the safest way to explore the area."
> —Sober traveler Rebecca Rush

Sober trip idea: Jordan

Most people visit Jordan to see the ancient city of Petra, which was famously carved from rose colored rock and served as the capital of the Nabataean Kingdom dating back to about 300 B.C. It's one of the world's most mind-blowing archeological sites. I could hardly believe it was real, even when I was there. But there's so much to see in this country. One of my favorite parts was staying in a canvas-sided tent cabin in a desert camp. You can get up at dawn and go for a camel ride. Or stay at the remote candle-lit Feynan Ecolodge within the Dana Biosphere Reserve and visit a Bedouin tent camp for coffee and a lesson in Bedouin culture. Since it's a Muslim country, alcohol is not readily available or part of everyday life. Sure, tourists can get a drink in hotels. But I loved how not in-your-face alcohol was here.

GENERAL SOBER TRAVEL TIPS

*T*he following tips apply to many sorts of trips, whether you're traveling for vacation, family events, or business.

• "First and foremost is having a way out," said recovery coach Sarah O'Brien. "Drive your own car. Park at the end of the driveway. Don't get blocked in."

• Rethink your favorite old activities. If vacation has always meant casinos and nightclubs, it's time to find new pleasures.

• If you attend recovery support groups, plan ahead. "Know the location you're going to," said O'Brien. Map the distance between where you're staying and the nearest recovery meetings. "Look up different recovery-based meetings in the area. If you get uncomfortable or uneasy, there's a place you can go."

• Let trusted people know what you're doing. Keep a friend on standby if you're going into a stressful situation or charged emotional event. A few little check-in texts can make a world of difference, as can just knowing someone has your back.

- "Don't go sit at a bar by yourself," said George Koob. "Sit at a table." Who needs a view of hundreds of shiny bottles? Don't feel guilty about taking up a whole table, even if you're on your own.

- At social occasions, keep a drink in your hand. Sparkling water with a slice of lime will do the job. People are less likely to offer you drinks when you already have one.

- Prepare to be patient. Travel is a lot more than just the Instagram moments. It's also delayed planes, boorish family members, and unexpected bad weather.

- Beware of the disinhibiting effects of social situations. Studies show that people can get intoxicated even if they're not drinking, a sort of contact high. While on the one hand this means you can loosen up without the alcohol, it's also possible to loosen up so much that you decide it's okay to drink.

- Avoid facing a stocked hotel room minibar by calling ahead and asking staff to remove all alcoholic beverages from your room. "Willpower is finite, and getting all the booze out of your room immediately ensures you won't get caught off guard with a negative emotion you can't handle—and a fridge full of booze you really can't," said sober traveler Rebecca Rush. If you arrive and the staff forgot to empty the minibar? Call the desk and "have them send someone up with a basket to clear your room of all that poison," said Rush. "It's incredibly empowering."

- Travel with other nondrinkers, if possible. "Going even with one other sober person makes it so much

easier," said Cole Bressler, founder of Choose Life
Sober Adventures, a travel company that specializes
in group trips. One of the reasons sober travel can
be so hard is the feeling of being the odd one out.

• Beware of all-inclusive vacations if you think the idea
of unlimited alcoholic drinks might be too tempting.

• If you do choose to visit an all-inclusive resort,
negotiate for a nondrinker rate. Sober coach Amanda
Kuda says most places have accommodated this
request. Since the biggest part of the all-inclusive
cost is usually for food and alcohol, she said,
"Usually people are willing to budge a little bit."

• Stick to your routine as much as possible. Eat regular
meals, meditate, pray, exercise—whatever you would
usually do to stay healthy and sane.

• "Sometimes I learn how to say 'no alcohol' in a local
language beforehand," said sober traveler Eric
Lucas. (See page 185 to learn how to say "I don't
drink alcohol" in 33 languages!)

• Focus more on day than night. "Whenever I travel
I typically don't stay out super late," said sober
traveler Nicole Kale. She avoids the late-night
drunken crew by having dinner earlier with the
families and old people and watching the sun set.
"Then head in and try to wake up and do more fun
things in the morning instead of staying out super
late." Nicole has promised to take me on a cemetery
tour when I finally make it to Charleston, SC, where
she lives.

Of course, some folks are night owls. Just because you don't drink doesn't mean you can't go out at night. A few ideas that don't center on alcohol: watch a performance, such as a play, stand-up comedy, music, or spoken word; visit a late-night bookstore; look for after-hours museum events; do something outdoorsy, like joining a moonlit paddle board or snowshoe expedition; try an escape room; take a ghost tour.

• Instead of slinking around with her club soda, hotel communications manager Dani Mathews boldly brings her own non-alcoholic wine to upscale restaurants. She's even been spared the corkage fee when the server saw her wine was non-alcoholic.

• Focus on what you're gaining, not what you're losing by being sober. "When traveling to traditionally boozy destinations, a lot of the appeal for many people is alcohol, and traveling sober means either not participating in those experiences or going and not having quite as much fun," said Jason, a New York-based publicist and sober traveler. "However, it also allows you to explore so much of a destination without having to nurse a hangover the next day. You can maximize your time and feel more in touch with the destination."

• On group trips, consider opting out of certain activities. "If it's not fun without alcohol, it's probably just not fun," said Amanda Kuda. "And therefore, you don't need to do it."

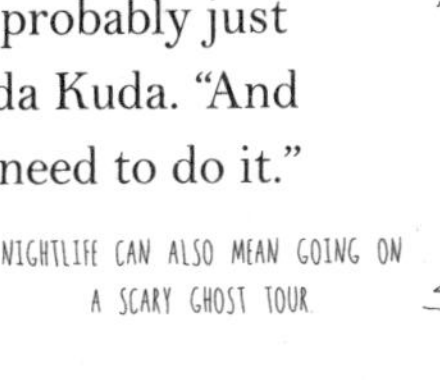
NIGHTLIFE CAN ALSO MEAN GOING ON
A SCARY GHOST TOUR.

Sober trip idea: Iceland

Iceland has plenty of outdoor activities and a fascinating culture. Sober traveler Sarah O'Brien visited Iceland with her father, cousin, and uncle. For O'Brien, the key to successful sober travel is to schedule activities in advance. "We did really great things. We drove around the country. We went and saw the glaciers, we went to Blue Lagoon (a geothermal spa)." I've been to Iceland, too, and enjoyed the thermal pools, waterfalls, cultural museums, and looking for fairies. Iceland ranked number one for best international sober travel destination in a study done by Lantana Recovery. While alcohol consumption is high, dependence is relatively low, and there are tons of sober-friendly activities.

SOBER IN AIRPORTS AND PLANES

IN THE AIRPORT

You get to the airport early, you're worried about flying, and there they are: airport bars. This can be a real hurdle for nervous flyers in recovery, especially when flights are delayed.

"There's bars on every corner in an airport," said sober coach and traveler Sarah O'Brien. "It's like a relaxing thing:; 'Let's have a drink to take the edge off.'" People in an airport aren't worried about drinking and driving, after all.

Sober traveler Maddie used to be a flight attendant and now works in aviation, so she is an expert on airports and airplanes. And on airport bars. "The airport for me was a huge trigger," she said. "Because I loved the airport bars. I would spend a lot of time in airports flying standby. So even when I wasn't working we'd be killing time in the airport, and the first place we would go is the airport bar." Maddie's Sphynx cat Phoebe inspired me to create Sober Cat.

Now that she doesn't drink, Maddie avoids arriving too early, to minimize waiting time in the airport. She brings a book and downloads some shows on her phone to entertain herself. She packs a snack, fills her water bottle, and buys a diet Coke. "Because I personally know that when I'm overly hungry, tired, or thirsty, that can be triggering." If she gets stuck in the airport for a long time during a layover or delay, she finds a restaurant and orders a real meal while she waits. "But I'll always sit at a table. I don't sit at the bar." Otherwise, she'll look for a quiet corner where she can plug in her phone.

My favorite airport activities include: seeing how many steps I can get in, searching for the restaurant with the best vegan food, getting a soy cappuccino, calling my husband, looking at knickknacks and earrings in airport shops, charging devices, and working on my laptop. Most airports now exhibit art and sometimes have museum-quality informational exhibits about the region, which provides a nice opportunity to stimulate brain cells, rather than kill them in the airport bar.

Other airport activities and amenities may include a chapel or yoga and meditation room, massage, massage chairs, or rocking chairs. Giant international airports sometimes have hotels inside the security gates. I've rented small rooms at airports in Delhi and Singapore, and spent a night in a bizarre sleeping pod in Dubai. These accommodations are usually expensive. But if you're on a 30-hour international journey—especially if you're somebody like me who can't sleep sitting up—after a while, sleep is worth an inflated price.

Airport SOS

Sober traveler Maddie shared a valuable tip. If you are freaking out in the airport and need sober support, talk to an airport service person. They can ask over the loudspeaker if there are any friends of Bill W. near your gate. (This stands for AA co-founder Bill Wilson.) "That's kind of a code for people in recovery, like, 'Hey, this person is struggling,'" said Maddie. If you're a nondrinker who hears this announcement, please heed the call and go help the person. Even if you abstain for other reasons rather than being in recovery, that person will be happy to see you. It feels so good to help somebody while traveling. I recently had the honor of talking down a stranger who was having a panic attack while we waited in a hot, sweaty mob at the Madrid airport. It transformed an irritating hassle into something meaningful.

ON THE PLANE

A little planning ahead can also help with successful sober flying. Maddie pre-selects her seat. "I'm really triggered by things being disrupted. My routine or whatever," she said. She likes to tuck herself away in a window seat where she can lean her head against the window and go to sleep.

She also brings a special snack for flying. "I'll go to the grocery store the day before and pick out a treat for myself. Usually it's chocolate. But a snack or something that I'm really looking forward to. That way I'm not even tempted to order something." After once having a slip and ordering a drink on a plane, she avoids all food and beverage service, only consuming what she brings on the plane with her.

We non-plane sleepers should bring something to occupy us, whether it's a book, a laptop, preloaded shows, a journal, a sketchbook, or games. Headphones are useful. While it's possible you'll meet somebody interesting on the plane, it's just as likely you'll want to signal to them that you're not interested in talking. Especially if they're drinking. You don't need their alcohol breath aimed at you in such close quarters. And they sure can start early! Recently, my seatmate on a two-hour flight that took off at 6 a.m. downed two gin and tonics.

Usually you're stuck with your seatmate. But if you feel really uncomfortable with a person's drinking, and there are empty seats on the plane, you can go to the rear of the plane where the flight attendants hang out and discreetly ask to change seats. Don't be ashamed—they've heard it all—but be very polite.

THE PERILS OF DRUNK FLYING

INSPIRATIONAL PLANE READING

One of the problems with drinking in airport bars is that it often results in drunk flying. We've all read in the news about horrifying examples of bad behavior on planes. Maybe we've seen it. Maybe we've done it. Even if you arrive at your destination without the police meeting you at the gate, it's not a good experience.

Maddie overdrank at airport bars a few times. "It was kind of scary because I didn't really remember the flight. Then you're in a whole new city and it's almost like, 'How did I get there?' Like I knew I took a flight, but I was also like, 'Oh my god, I was on a plane interacting with people and I don't remember it.' And then also being out of control of myself in a place I don't know where I don't know people, that scares me, too."

While she never drank on the job, as a flight crew member she drank on layovers. "We would have these long layovers in towns that I didn't know what to do. And it's just kind of frightening looking back on it now to be like, I was this young girl by myself in these weird cities. And yeah, there's times I was heavily intoxicated alone. And I'm really grateful that nothing bad happened to me, just looking back and realizing how out of control I was and didn't know it. Especially in a place where nobody would have known who I was or helped me if I needed it."

> "Air travel is dehydrating and exhausting. Drink more water than you want to. It's okay if you get up to pee every 20 minutes. I always get an aisle seat for that. Also aisle yoga. It looks ridiculous but completely changes how you feel upon arrival."
>
> —Sober traveler Rebecca Rush

Sober trip idea: Barcelona

Lane Kennedy, a mindfulness coach and self-described "self-preservation specialist," loves Barcelona, Spain. She recommends visiting the Sagrada Família church designed by famous architect Antoni Gaudí. "Gaudí's basilica isn't just architecture; it's a profoundly spiritual space," she told me in an email. "Visit early in the morning for tranquility and the colorful light filtering through the stained glass—it resembles a meditative portal, which is a bonus in my mind. You can certainly skip the audio guide and sit in silence inside. Allow the light to speak for itself." She also recommends walking on the beach and eating at Tapas 24, one of the city's most popular tapas bars. "There are great 12-step meetings around the city that are easy to get to from anywhere." Some are even in English.

TRAVELING WITH DRINKERS

Sober travel is a simple and enjoyable concept: Seeing the world without using a substance that, after a brief high, has adverse health effects and can get you into all sorts of trouble. But like so many other things in life, other people can foil a simple aim. While this book is meant to celebrate all the things we can do when we choose to live life unanesthetized, I found myself writing a list of times I've gotten pissed off when traveling because of other people's drinking. I'm including them here because I'm sure other non-drinkers can relate, and because I don't want to gloss over the hard parts of traveling sober. Many of these come from my work as a travel writer. Professionals in other industries have surely experienced parallel scenarios.

Times I've gotten pissed off by other people when traveling sober:

- People acting like I'm missing out because I don't drink

- PR people endlessly promoting their wine regions. When did every place in the world start growing grapes?

- Getting dragged along on a tour of a brewery, winery, cidery, or moonshine plant

- Stuck at a dinner table for hours while people had
  drink after drink, endlessly commenting on said
  drinks

- Hiding out in the bathroom of a trendy restaurant and
  calling my nondrinking husband for support because
  our group of travel writers had a surprise mixology
  class

- Feeling panicky trying to network at cocktail parties
  and like I was blowing business opportunities, then
  feeling bad about myself for ducking out early. I've
  been sober for ages; shouldn't I be able to handle
  this?

- Feeling unpopular and unlovable and priggish and
  lonely because I couldn't have a few drinks with the
  group

Obviously I'm a little resentful and defensive about drinkers even after all this time. Doing research for this book and meeting a wider variety of non-drinkers has helped shift my attitude a little.

I approached Ernest White II, host of the PBS travel series *Fly Brother,* after hearing him give a keynote speech at a travel industry conference. I shyly told him about my book and was surprised when he said, "Oh, I'm a sober traveler, too!" After further conversation, I learned he never had an alcohol problem. He simply doesn't like the way

it makes him feel, so he doesn't drink. That was the first time I realized that nondrinking people who aren't in recovery might also see themselves as sober travelers. Revelation time!

"I'm the only person I know quite often who does not drink alcohol," he told me. "I'm very much around plenty of friends and family members who will have a little drinky poo." He manages to be firm in his decisions without judging others. And he avoids places where the focus is on alcohol. "I don't shut it out of my life. It's not like I've created a life around a bunch of other sober folks who look down on alcohol and all that kind of stuff. That's not really my life either. It's just not my choice, and therefore that has served me. And it continues to serve me."

The next few sections of this book deal with some of the situations you may find yourself in when you travel with drinkers. It's about setting boundaries to keep yourself safe while being open enough to experience the people and events around you. And also about assessing whether the event is worth going to or not. Do the benefits outweigh the risks to your sobriety and serenity?

EXPLAINING YOUR SOBRIETY TO OTHERS

If you are traveling with people who drink—say you're on a group cycling tour or a cruise—sooner or later somebody will probably ask about your abstinence. Remember that it's up to you how much you share. You don't owe anybody a thorough explanation of your life's history and choices. Then again, if you feel like telling them, that's fine, too.

This topic brings up a lot of emotion in some nondrinkers. For people in recovery, the decision not to drink is life or death. For people who don't drink for other health reasons, this question is just as offensive.

I once found myself sharing a villa in the Virgin Islands for a destination wedding with five heavy-drinking guys from Louisiana. Within an hour of meeting them, I told them I didn't drink, to which one said, "Oh, you'll be drinking by the end of the weekend." I'm sure he just meant to include me in the fun, but I responded with all the grace of a cornered raccoon.

Here are a few things you could say when somebody offers you a drink, then persists beyond your polite "No thanks":

- "No, thanks. I'm taking a break."

- "No, thanks. I don't like drinking alcohol." (Although they might respond, "Oh, but I know you'll like this drink!")

- "No, thanks. Alcohol doesn't agree with me."

- "No, thanks. When I travel, alcohol just screws me up."

- "No, thanks. I'm allergic to alcohol. I break out in handcuffs."

- "No, thanks. Abstinence is part of my spiritual practice."

- "No, thanks. I took a temperance oath."

- "No, thanks. When I drink, I behave very badly." (Though some ass might laugh and say, "I'd love to see that!")

- "No, thanks. Alcohol interacts with my anti-psychotic drugs and I can't be accountable for my actions."

What's the common denominator here? *No.* The rest is up to you. And if people persist in offering you alcohol after you've made it very clear that you don't drink, that reveals more about them than you. They're probably uncomfortable with their own relationship with alcohol and want everyone to join them in a pint or ten to normalize their drinking behavior.

The "why don't you drink" conversation is an area that can stir anger and defensiveness in recovering alcoholics. But again, this is a good time to take a cue from non-recovering sober travelers who can say no without feeling the same charge. "I've gotten enough practice just to set a boundary in a way that's still respectful to my hosts," said Ernest White II. What does he say when he's offered a drink? "No thank you. No thank you. No thank you. No thank you. No thank you," he told me, laughing. "And I just keep saying, 'No thank you.' That's all. 'Oh, but it's…' 'Well, no thank you. I'm fine. No thank you. I appreciate it. No thank you.' Graciously declining. And consistently." And if somebody really doesn't like that he's declining a drink? "If it is taken disrespectfully, well, there's nothing I can do about that. I'm not going to start drinking alcohol in order to make people feel better."

You can also get creative in your excuses. A friend put me in touch with sober traveler Matt McDermott, who is in the construction field in Illinois. A lot of his coworkers like to drink. "Every year, at the company Christmas party, there was a lot of pressure. People would remind me that I worked really hard and could cut loose for one night. I ended up lying! I would wear my coaching shirt (I coach youth wrestling) and say I needed to be at an event that night." He sometimes used a diversion tactic when people asked him about his nondrinking. "Sometimes, I would either pretend to not hear the question or answer a totally different question. Example: Why don't you drink? I'd answer the question, 'What was your first construction job?' I'd go into detail about my first job, then the second. Tons of detail.

Eventually, they would move on to a different subject. Now, I just tell people I'm in recovery and I had enough to drink in my twenties to last a lifetime."

Everybody I interviewed for this project has encountered pushy people who question their nondrinking. Here are a few of their experiences.

"Occasionally at an academic conference, a colleague will make a snarky remark," said public historian Jason Steinhauer, who I know from my work in the oral history field. (Many of us writers have multiple careers.) "This tends to happen in Europe, where drinking is a staple of university life. Once I explain I am not drinking for health reasons, the snarkiness usually ends." He doesn't let it get him down. "Be confident in your choices. Don't be self-conscious. Your health and peace of mind are always the most important considerations."

Sober travel writer Eric Lucas (who is also a hay farmer—see what I mean about multiple careers?) tells people, "I don't drink. No, not at all. Period." Even after this definitive declaration, some bozos persist. "I have heard some outrageously stupid remarks, such as, 'How do you have any fun?' or 'I can't believe you're a writer and you don't drink!' I respond by expressing amazement at their ignorance."

"My go-to phrase is 'I don't drink,' which is harder to say than you'd think," said Jason, a publicist in New York. "I always want to qualify it by saying, 'I don't *really* drink' or 'I don't drink that much,' but both of those phrases leave room for questioning. I personally think it's very rude for people to ask why. It's none of anyone's business and it can be a painful subject for a lot of people. But I'm almost always asked why, and I usually respond that I don't drink for health reasons and because in general, I just don't like it."

"I always tell people I'm driving, whether I am or not," said sober travel writer Fred Wright. "In situations where it's obvious I'm not traveling by car (like in foreign destinations), I just decline politely and thank the person for the offer."

"I simply say, 'No, thank you,'" said author Candy Arrington, whom I met at a writing conference in New Jersey. "If I'm pressured about why I don't drink, I tell people I grew up in a family of nondrinkers and made the decision in college to continue following that path because I have seen how drinking ruins lives. Alternatively, I say not drinking is a personal conviction based on my faith."

"When people would ask, I'd say, 'It wasn't fun anymore. I didn't feel good and it wasn't fun,'" said Brandy Cogsdill of Greenville, South Carolina, referring to the days she worked in corporate finance. Now she's a partner in the excellent Southern Pressed Juicery, and the people that buy her detox juices probably don't need to ask her this question.

> "If pushed, my response is typically, 'I believe that people have a limit of how much they should drink in a lifetime, and I already hit mine.' It usually makes them wonder if they've hit theirs!"
> —Anonymous in Montana

Since I started researching this book, I'm much more upfront about being a sober traveler. While I never hid my sobriety, I seldom led with it. Now I tell people that I write about sober travel and I try to say it like it's a perfectly normal subject. Unfortunately, it's still not always well received. I was at a session about accessible travel at a large travel conference. After listening to people talk about how they managed to travel in wheelchairs or with other physical challenges, I told the woman beside me that I focus on sober travel. She said, "I can't imagine traveling without drinking." We'd just spent an hour listening to people talk about their determination to get around the world without being able to use their legs. Would she go up to them and casually tell them she couldn't imagine traveling without walking? And is it really easier for people to imagine giving up walking than drinking?

As travel and outdoors writer Lisa Maloney pointed out, "I think that until we start normalizing the plain and simple 'I just don't want to drink' as a valid reason for not drinking, we're actually perpetuating the societal norm of drinking as the default."

Not up for negotiation

Sober coach Amanda Kuda said nobody has pressured her to drink for years because she confidently communicates "no" from the beginning. She encourages her clients to make a little script and practice saying no ahead of time. "I think that as you gain confidence by keeping commitments to yourself and flexing your sober muscles, you get to a point where you can say things more confidently."

Your nondrinking cannot be negotiable. "Because when it's negotiable you become instantly weak. And every time I've traveled, I've gone into it saying, 'I'm not drinking on this trip.' And I did that for a year. And now I just don't have to make that negotiation with myself."

Amanda also brought up the phenomenon of nondrinkers explaining their choice even when no one asked. "We get so anxious because there's such a social stigma about not drinking that we often verbal vomit and over-offer." Amanda urges you to spare listeners your life story and keep your dignity with a simple "No, thank you."

Sober trip idea: Ureka, Equatorial Guinea

You want a real adventure? "Equatorial Guinea is a small country but at the same time, one of the most interesting countries in the world," said Héctor Nguema, founder of Rumbo Malabo Travel Agency. Héctor contacted me after I solicited ideas for sober trips on a website that matched journalists with sources. The town of Ureka stands out for the difficulty of getting there, being the wettest place in Africa, and for its beautiful beaches, he told me. "In order to access them, a permit is needed in the Ministry of Tourism in addition to having the visa to enter the country, then drive for two hours and ascend a mountain of more than 2,000 meters high. In the city of Ureka we also have a beautiful camp, called Moabá, in which one of the rules to follow is the strict prohibition of carrying alcohol or directly consuming it." There's also a beautiful waterfall.

BIZ NETWORKING WITHOUT THE BOOZE

*A*re all the deals made at the bars? As a writer, I've felt like I've lost out on so many biz opportunities because I'm not staying up late drinking with editors. People in other fields have shared similar worries. Without the lubrication of alcohol, we have to approach networking in a different way.

Generally, business conferences and events are not the place for spilling your whole personal story about why you don't drink. It's not the best use of your networking time. And, unfortunately, there's still a stigma around alcohol abuse disorder. If you don't drink because of other health reasons or personal convictions, the revelers in the group might not like that, either.

When pressed to drink by your business acquaintances, NIAAA director George Koob suggests offering a business excuse. "You know, 'If I drink, I'm not going to be in top form for our meeting.' Or, 'I need to get a good night's sleep.'" These excuses are true without having to get into your personal details. Dr. Koob acknowledges that it can be tough when pressured to drink by your coworkers. "You just have to say no. I hope you don't have to get to the point where you have to surreptitiously pour the drink into a plant."

Sober coach Amanda Kuda stresses to her clients that they have two main options when confronted with a business networking event that they don't really want to go to. One is opting out. "You have to draw a line between what is required networking and what are you making up in your mind that you need to do," she said. So try to assess how useful the event might be. Amanda finds it hard to connect at big, rowdy venues, so unless she feels the event is absolutely necessary, she'll skip these and attend smaller, quieter events.

"The second option is, you can go. You can always go and test the waters. See what you're able to tolerate. Maybe surprise yourself and find out that something is more fun than you expected it to be. And if you choose to go, you can always leave. You don't have to stay the full time," Amanda said.

I've been to a ton of business events. The ones I personally dread most are the cocktail parties where everybody stands with a drink, just staring at each other. It's loud and there aren't many seats. Sometimes I last 15 minutes before fleeing. But other times, I've stayed till the end and shut the party down. The difference? A blend of my attitude, the other people there, and the activities. It makes a huge difference to me if there are things to do. Sometimes event planners include games like ping-pong or cornhole, photo booths, or exhibits to look at. Just standing around trying not to stare at everybody else's drink is the worst.

Does drinking really help you get ahead in business?

Depending on the industry in which you work, not drinking may or may not hold you back from opportunities. Brandy Cogsdill said she missed out on relationship building with peers by not drinking. "Because if everyone's going out, and depending on your season in life, if that's not something you can do, you can miss out on that exposure." However, she reports, the bosses weren't partaking. "As you get more to the management levels or executive levels, most of the people I worked with were not big drinkers." Peers drank, but company owners were less likely to. "I don't think that not being a drinker ever held me back," she said. She worked in finance, and who wants heavy drinkers looking after large sums of money?

REMEMBER, YOU'RE A PROFESSIONAL

One thing I found interesting in my research is that as a person in recovery, I'm more self-conscious about not drinking. Some of the nondrinkers I interviewed who refrain from alcohol for other reasons were much more matter-of-fact. It's a less charged situation for them. As senior salesperson and sober traveler Emma Orr put it, "Should we really be drinking when we're working?"

Mira T., a Chicago-based travel writer, avoids alcohol for health reasons and because she's never much liked it. "I'm not much of a party girl," she said. "On a press trip, I don't go out on those late-night pub crawls with the group. I take great pride in being a professional on a press trip and I do not want to be pegged as a 'drinker with a hangover.'" Many people find it hard to handle themselves around free drinks, she noted, and this does not benefit their professional reputation. "In a group, you quickly find out who's there for the free alcohol when they order three or more drinks. It taints you." I have heard travel writers gossip about mutual acquaintances getting embarrassingly drunk at industry events. Drinking is generally acceptable, but only if you can hold your booze without obvious consequences.

One time I was at a travel writing conference in Coronado, California. We were sitting on the beach making s'mores. I didn't realize until we got up to leave that one of my fellow writers had put away way too much wine. She could barely stand. I had never met her before that weekend, but I immediately felt for her. I helped her discreetly get back on the bus that took us to our hotel. She kept dropping her keys and stumbling against the walls of the hotel corridor. I had a visceral memory of how hard it was for me to hold onto keys when I was drinking. There were times I curled up on my doorstep in New Orleans because I'd lost my key. At the end of my drinking, I wore my house key on a cord around my neck to avoid losing it. So I made sure I got this woman and her key safely inside her room. I told no other writers about it. That old saying "There but for the grace of God go I" is all too true. Instead of gossiping about people's drinking, I want to support those who need help opting out.

While many people consider alcohol an essential sober lubricant, if you're reading this book it's likely you know how easy it is to go a little too far and make a complete ass out of yourself. Sober networking is so much safer. "Even if you are a little anxious, you're going to have full command of your body, of your mind, of your words," said Amanda Kuda. "If you're actually responsible for making connections at work, you are going to be so much better off in the long run doing that sober."

NOT DRINKING CAN BE UNCOMFORTABLE

I don't mean to make it all sound easy. Not only might we have an innate desire to drink, but sometimes we face real or perceived pressure from others. "We do seem to be an alcohol/drug/pill-fixated society, and that can make the rest of us feel a little like outsiders," nondrinker Elizabeth Harryman Lasley told me.

"It has been hard at work when there is champagne and we are celebrating, and I have to say 'not for me,'" said Emma Orr. "No one wants to be the odd one out."

Brandy Cogsdill described her experiences at corporate finance conferences with the same language. "Often you're viewed differently if everyone's ordering drinks and you're like, 'No, thanks' or 'I just want a water.' 'Why?' 'I don't drink.' It's like you're the odd person out, rather than the person who may be intoxicated or acting inappropriately. You're the one that everyone looks at differently."

There's the real chance that you might be excluded, though some sources feel this is getting better. One sober traveler who wished to remain anonymous worked as a wildland firefighter for 20 years. "When I first quit drinking, I'd be asked out with the crew to do things," she said. "But after they saw I don't drink or smoke pot, I wouldn't be asked again." Fortunately, she experienced more acceptance and inclusivity later in her career.

Another anonymous source, a London-based travel writer, bemoans the place of alcohol within his industry. "It is often the case that networking events in travel are centered around or at events where alcohol consumption is actively encouraged and people frown upon (publicly and privately) those that refrain or appear uncomfortable with this," he told me. "I am often slighted for potentially not being as 'fun'—I am very transparent about being Muslim and someone who does not drink alcohol." His declaration of not drinking "inevitably leads to being the 'party pooper.'"

For her part, Brandy experienced unwanted attention just for ordering water when out with coworkers. "It became a joke," she told me. "'Oh, it's Brandy. She's the most fun person here.'"

Author and sober traveler Candy Arrington faced direct criticism for not drinking. "When I was younger, I worked for corporate attorneys who often entertained clients. My not drinking at these events flagged me as non-compliant. One of the attorneys pulled me aside and told me I was making the clients uncomfortable. I told him drinking made me uncomfortable and I wasn't going to drink because it was a personal conviction. I was surprised he pushed the issue, but I didn't lose my job because I wouldn't drink." In her later career as an author, Candy has sometimes felt shunned at writing conferences. "Several people who are big drinkers have made unkind comments, treating me as if I am trying to be 'too good' and excluding me when they go out to eat or gather for conversation. I don't make a big deal about not drinking, and I don't appreciate that they do."

Then again, we might be making ourselves needlessly uncomfortable when not drinking really isn't an issue for our higher-ups. Sober traveler Mary Paige Rose, who I met when we were in an art group together, once worked herself into a tizzy for nothing. "When I was the marketing consultant for a Japanese company, we were at the closing dinner for the project. The president of the company had brought from Japan two bottles of saké that a Buddhist monk had blessed for the dinner. I was really stressed about it because everyone was expected to drink the saké for a toast to the project. I had a moment of clarity and when he came to my glass, I just put my hand over my glass and he passed me. No big deal except in my head!"

Sometimes it just feels lonely not to drink. Writers are notorious for drinking heavily. I've been to many writing conferences where the hotel bar was packed. But when I went to a healthcare writers' conference, I was surprised that after an afternoon with such grim sessions as understanding the opioid epidemic, the participants were gung-ho to attend cocktail hour. It was as if there was a divide between poor addicted souls as subject matter and a fellow writer like me, in recovery, sitting right next to them. And for God's sake, they weren't Hemingway or Bukowski—they were writing about healthcare policy! *Whatever,* as the serene say.

So, external pressure to drink is on a spectrum. Why do some people want you to drink more than others? It probably reflects their issues with alcohol. Hold steady. No matter what somebody else thinks about your not drinking, you are the one taking yourself home at the end of the night.

I was at a huge travel industry event in Los Angeles. One night we had the option to take shuttle buses to the Santa Monica Pier and spend the evening riding rides and playing arcade games. I broke away from the industry flock because I wanted to use my unlimited ride bracelet. While I was on a ride called the Pacific Plunge where you go straight up high, then drop fast, I noticed one young woman who was standing and watching but not getting on. The ride operator and I encouraged her, but she was scared of heights and the drop. Finally she got on and sat next to me. After that, we just started riding rides together. She was a Black woman from Florida studying to be a doctor who bought herself a solo trip to L.A. for her twentieth birthday. I was a white West Coast writer in my fifties. There was no apparent reason we'd befriend each other, but we spent the next two hours riding all the rides and playing arcade games. She beat me at a water gun game and won a green triceratops plushie. We played Skee-Ball and took photos of each other on the pier with the sunset. When we parted ways, we both said we'd been having fun alone, but that it was more fun together. She'd been missing her best friend, and I'd missed my sisters.

I rejoined the industry people on the shuttle bus. "How was your time on the pier?" I asked the travel professional who sat next to me "Great!" he said. I asked, "What did you do? What rides did you go on?" He seemed a little perplexed and said, "We just hung out at the bar." Then he passed out beside me. Had I really missed out on anything by not drinking?

Sober trip idea: Quebec City
My favorite way to see Quebec City is through a stay at Le Monastère des Augustines. In the early 2000s, the order of Augustinian nuns confronted their dwindling population and converted their huge 14th century monastery and hospital into a wellness hotel. It opened to the public in 2015. They left lots of historic touches—Catholic statues and small monastic rooms—but added yoga and deluxe bedding. As a Catholic/yoga/12-step person, I feel like this place was made just for me. The monastery is in the original walled city of Quebec, so you can wander around all day, museum hopping, people watching, shopping, and going to restaurants, but come back anytime to what feels like a very safe and hallowed space. Or you can stay in and participate in yoga and tai chi classes and get a massage. The onsite spa services are aimed at improving your health, not your vanity. They even have excellent vegan food. I love that the monastery still maintains the tradition of taking breakfast in silence, like the nuns did. Who needs chit-chat that early?

PRACTICAL SOBER NETWORKING TIPS

How do you network without alcohol? For people who have quit drinking, this can be a mystery. So let's consult Ernest White II, who travels around the world meeting people for his Fly Brother PBS travel show. "As someone who's never drank alcohol with regularity, I had to develop the skills of networking and engaging without that crutch." He recommends smiling at people, making eye contact, and nodding. This sounds obvious when you read it, but how many of us have spent time at events anxiously hiding in a corner and staring at our phones?

"Recognize other people's humanity. Recognize people in the room." He often starts a conversation by complimenting people. "If someone's wearing a nice watch, I'm going to say they have a nice watch. If their hair looks nice, no matter what their gender expression is, I'm going to say like, 'Oh my God, your hair looks amazing.'"

When feeling intimidated by someone who is very beautiful or accomplished or whatever, Ernest says he brings to mind one of his favorite book titles: *Everybody Poops*. "I think that's funny and true and it's a great way of stepping out of the insecurity of engaging with other people. Because everybody poops. So there's no need to feel intimidated unduly."

Many people I interviewed stressed the usefulness of always having a drink in your hand. "If you get there early, order yourself a drink prior to your colleagues or other people arriving," advised sober coach Sarah O'Brien. "That way people aren't tempted to ask you, 'Hey, what would you like to drink? What would you like from the bar? What can I get you? Are you going to have another?' Always just having a drink in your hand can make for that uncomfortableness to kind of disappear."

Keep track of your drink—after all, your ginger ale might resemble somebody else's whiskey and soda. Hold onto it or know right where it is. This isn't super elegant, but sometimes I drink from a clearly labeled can so I don't accidentally pick up a mixed drink.

"At many cocktail events and meals, I find that a non-alcoholic beverage is rarely available beyond sodas, fruit juices, or water. It's assumed that you'll drink a glass of wine or the crafted cocktail of the evening," said sober traveler Mary Chong. "When I'm there with my glass of soda or water, the host/fellow guests often question why I don't have a 'proper' drink—which is quite honestly uncomfortable and awkward. I truly wish that hosts were more cognizant of this and would create a crafted non-alcoholic 'festive' drink versus it being an afterthought."

Mary—who doesn't drink for health reasons—sometimes orders half a glass of wine. "I've just held it/left it on the table to avoid any questions. I've also asked the bartender/server to put my soda in a cocktail glass and to make it look festive." While I would not recommend that recovering alcoholics attempt to hold onto a glass of wine without drinking it, this is a possible approach for people who don't drink for other reasons.

Offer special non-alcoholic drinks at your events

If you happen to be planning an event, for every "signature cocktail," introduce a signature mocktail. I was once at an event at Fountainhead Antique Car Museum in Fairbanks, Alaska and was blown away that they had dispensers of two different special non-alc drinks. I asked Fountainhead's food and beverage manager Carly Nelson about this and she said, "We definitely promote offering a mocktail service to our events, as I know how important it is to have delicious options for all to enjoy. I absolutely love creating mocktails that are more unique to create the same buzz and excitement that guests also experience ordering a regular cocktail. I can't count the number of times I'll go to a bar and ask for a mocktail and am just offered juice with Sprite on top. I thrive on creating a better option for those who choose to not drink or perhaps be a designated driver for the evening." I love to hear this.

Sober traveler Maddie carefully considers which events to attend. "If it's just drinks, I usually don't go. But if I know there's food there, I'll go and order an appetizer and sit. But

I don't stay long. And I always drive myself so that if I feel uncomfortable or weird I can leave."

What if you arrive at an event and it's a total booze fest? "I wouldn't go to a reception that didn't have alternative beverages," said NIAAA director George Koob. "If you have to actually beg to get a sparkling water, I would turn around and leave."

Sober traveler Sarah—one of my friends since junior high school who also quit drinking—notes that alcohol plays an important role in business because it brings people together. They network in insanely long bar lines at receptions. "And for me, I actually have to make the effort to approach people, to move around the room. My purpose of movement isn't to get to the front of the line. My purpose for movement around is to actually connect with people, network with people. So it's scary. It's scary. I don't have a buffer like that. I get to approach people because I actually am interested in learning about them."

"'Let's grab a drink' is probably the most common networking initiating phrase or scenario," said Jason, a New York–based publicist and sober traveler. "I've felt pressured to imbibe on more occasions than I can count, and often have to not only explain that I don't drink but also explain why, which always feels a little rude to me. My tip would be to ask for a coffee meeting instead during the day—much less likely to lead to alcohol." Or explain up front that you'd like to grab a non-alcoholic drink to squash any expectations, he suggested.

However, Sarah, who writes and edits best-selling inspirational business books, looks at not drinking as a gift. "I could feel very shut out. And at the same time, I'm so grateful to be shut out. I'm not interested." Not drinking gives her the freedom of choosing when to participate and when to go back to her hotel room on her own. "I have the freedom of choice because I'm not beholden to getting myself more alcohol. So I don't have to go." Instead of a hangover, she wakes up fresh for business meetings. "Because I love work. Work to me is my love made visible. And what I'm doing, I get to make a contribution. And when I drink, I don't get to make that contribution."

FIND YOUR SOBER BIZ NETWORKING ALLIES

What makes an event ten times easier? Sober friends and allies. Notice who else orders soda water or turns their wine glass upside down at a seated event. Prick up your ears when somebody else says "I don't drink" and gravitate in their direction.

"I have found that if I share I'm not drinking, there's always somebody else who's like, 'Oh, I don't drink, either,'" said sober traveler Maddie. "And I'll sit with that person. So I've actually made several friends at my job who don't drink. If we go, we always sit together." They may or may not explain why they skip alcohol. "I shared a little bit with my one friend," Maddie said, "and he's shared that his is like mental health and medication reasoning. And we're just like, cool, yeah, not a big deal."

Author Candy Arrington has also found her people at writing events. "Over the years, I have connected with those who don't drink at the conference, and that has made me feel less left out. You begin to gain a sense of who values networking that doesn't involve drinking. They also are looking for non-drinkers and gravitate toward you."

Another plus: If you go out to dinner with your nondrinking colleagues, there's less potential hassle about the bill. "I don't feel pressure to drink because I don't give two shits what people think in that regard," said sober therapist Sugar. "But I have felt pressure to pay extra for other people's alcohol consumption, which is really unfair. It's not such a big issue these days since I'm more financially secure, but it was really challenging for a number of years." Maybe people will think I'm tacky, but I would ask for my own bill in this situation.

Sometimes vulnerability pays off

Tamar Medford, founder of the podcast *You're Sober! Now What?* quit drinking while working in sales in the marine industry. People in the company she worked for knew she had a problem. "So they were really proud when I quit. But in the industry, that's still the lifestyle oftentimes." Initially it was hard to go to industry events as a sober person. She either felt like she had to hide or explain her nondrinking.

"And of course I didn't like being around drunk people. But the longer I've stayed sober, and after basically coming out with 'Hey, I'm in recovery and starting a podcast,' it became easier to have those heart-to-heart conversations with customers. Because often me not drinking would have them go, 'Oh, you're not having a drink,' and I'd say, 'No, I'm a retired professional.' And they'd laugh. But always, inevitably, it opened up the conversation to them saying, 'Oh, I've felt like I have a problem,' or 'My brother's in AA.' I find that the more courage I had to open up about it, the better my relationships got with the people that I worked with."

Sober trip idea: Berlin

Sobriety is on the rise in this city known for excess and all-hours drinking. In 2023, about one in five Berliners reported that they don't drink at all.[13] Berliner Gideon Bellin started his series of Sober Sensations parties in 2016, and has thrown more than 40 around Europe since then. These events start early and end around midnight, and focus on stimulating all five senses with high-production-value music, creative mocktails, natural scenes, light shows, and smoke and bubble machines. They cultivate a friendly atmosphere in clean, non-grungy venues. Supposedly even the door people are welcoming. Berlin also has alcohol-free spirit stores and non-alcoholic beer manufacturers. Germany's first alcohol-free bar, Zeroliq, has closed, but by the time you read this maybe someone else will have opened a new one.

---

13 "Sobriety on the rise: Is Berlin over the influence?" www.the-berliner.com/berlin/sobriety-on-the-rise-is-berlin-over-the-influence/ accessed May 14, 2024

FAMILY GATHERINGS

*D*epending on your family, get-togethers might be a time when everybody is expected to drink. If you're shaky in your sobriety—or if your family turns into scary cretins when under the influence—you might want to RSVP no thanks. But if you're feeling solid and basically get along with your family, you can use the same strategies listed for other situations and keep yourself sober while maintaining precious relationships.

Phillip Vitela grew up in a predominantly Latino town in California. As he explained in an article for The Sober Curator website, "In the Latino culture in which I was raised, admitting you are powerless over anything, especially alcohol, is an admission of weakness. We don't do weak; we don't admit to uncontrollable problems to our friends and family. I was taught to manage my own problems—this is being a man."[14] This description intrigued me. I understand this feeling, and have seen it play out before me, especially with men who think they are so tough and have it together when anybody else can see they're drowning in alcohol and drugs. Phillip, who now lives in Washington state, let me interview him on the phone about what it was like to get sober and then get together with his family.

---

14 "Being Brown in the Whitewashed World of Recovery" thesobercurator. com/being-brown-in-the-whitewashed-world-of-recovery/ accessed June 23, 2024

His uncles—and many families in Watsonville, California, where he grew up—worked for the local Budweiser distributor. "We all grew up in a culture that on Sundays we all got together as a family," Phillip said. "And on Sundays, they drink copious amounts of beer. We never saw any violence. We never saw any arguments. It was always a happy gathering, but tons of beer. And as my uncles and cousins got older, health issues develop, right? But it didn't really stop the alcohol consumption because it's part of what we do. And it wasn't just our family. It was everybody in our town."

When Phillip quit drinking, his family and friends were puzzled. But he was steadfast, and after about two years everybody got used to his not drinking. "The Latino community has a huge diabetes problem," Phillip said. "Alcohol really, really affects diabetes. So my goal is to lead from the front with my friends and relatives. You affect your own *oikos*, your own circle. And that circle spreads to another circle spreads to another circle. And it's kind of working. I have friends calling me up randomly. You know, 'I have an issue. How do I get out of it? What should I do?' Which is great for me. I love it."

SOBER AND JOYOUS AT A WEDDING

How many stories have you heard, or movies have you seen, about people getting heinously drunk at weddings? Perhaps it's even happened to you. What is it about these events that make relapsing so tempting? Let's look at a few factors. You'll notice they all have to do with feelings.

- Wanting to join in the celebration

- Being envious because you're not getting married

- Being sad because you're thinking of your own unhappy marriage, recent breakup, or death of a partner

- Feeling jealous because you're at the wedding of an ex you haven't gotten over

- Social anxiety because you hardly know anybody

- Social anxiety because you think everybody dislikes you

- Working up the courage to seduce a bridesmaid, groomsman, or wedding guest

- Deep-seated resentments against family members or other guests present

- Desperate greed because booze is free and you know your thirst for it is bottomless and the void within you can accommodate any amount

Such an emotionally charged scenario calls for finding an ally ahead of time. "My number one piece of advice would be to reach out to someone who will be there, who can be an advocate for you," said sober coach Amanda Kuda. "Tell them, 'Hey, I'm not drinking right now and I have to tell you I'm a little nervous about the event. I don't expect you to be my watchdog. But if you can at least be kind of like a cheerleader for me and support me or be a shoulder to lean on, I would really appreciate that. Can I count on you?'" Somebody who loves and cares about you will probably not say no. In Amanda's experience, people often hesitate to ask for help ahead of time, which sometimes leads to failure. But if you have somebody who is supporting you and holding you accountable, it can make all the difference.

Sober therapist Sugar takes this idea a step farther. "I've found it helpful to ask one other person to also not drink. It's something you have to arrange in advance and make sure the

other person is honest with you and themself that they won't mind so that you don't have to deal with resentment later." If she doesn't have a sober ally at an event, she's likely to only stay for a short time.

Like with all events, having an exit strategy is vital. Drive your own car, have your ride share app ready to go, or consult the local bus schedule. If you're traveling to an out-of-town wedding, try to get your own hotel room or share with another nondrinker so you have a safe haven.

Despite your worries, you might surprise yourself by having a fabulous time. Dani Mathews went to a wedding when she'd only been alcohol-free for a month. "I drank water, and though I was getting weird looks, I had the best time and actually felt more aware than at any other wedding. I was able to really enjoy it and dance the night away."

Whether your personal experience turns out to be awesome or just tolerable, remember that the point of attending the wedding is to support the couple getting married. This means showing up, looking nice, and sincerely congratulating them. Maybe bring a gift. Appear in some photos to prove you were there. While we all experience our individual feelings, our own emotions are not the point of the wedding. If you're feeling stable enough to go, put in an appearance, smile, and wish the couple the best. Maybe you'll dance all night like Dani. Or maybe you'll skip out early. Either way is fine.

Destination weddings

Have you ever attended a destination wedding? This could mean group camping in the woods or staying in a villa in the Virgin Islands for a few days. It usually involves a lot of togetherness with the other wedding guests and, unless it's a sober group, alcohol. Depending on your personality and mindset, this could be what priceless memories are made of, or an indescribable form of hell. Before you commit, try to get a feel for what sorts of activities are planned, who's going, and whether you'll have your own space. These types of events often require using your vacation days and spending a lot of money, so factor that into your decision, too. And tell the planners ahead of time that you don't drink so they can be sure to supply delicious non-alcoholic refreshments.

"Jewish weddings have a lot of alcohol, as do Jewish holidays like Passover and Purim. In these instances, grape juice is your best friend."
—Nondrinking traveler Jason Steinhauer

Planning your own wedding

What if you are the person getting married? Do you have to supply alcohol at your event? Do you want to? It's totally up to the couple.

When my now-husband and I were planning our elopement and small reception a few years back, I was adamant about not serving alcohol. Although he is also sober, he thought we should have alcohol available for those who wanted it. This was the one place I might have gone a little bridezilla. I said that if people couldn't spend an hour or two without a drink in my mom's backyard—the site of the reception—then fine, who needs them? They could go out to a bar instead and not darken Mom's doorstep. There was no way I was buying a bunch of booze and watching other people drink it. Fortunately, he relented and we managed to proceed.

If you're in recovery and most of the guests are sober, I think you absolutely should keep things alcohol-free. If any of your guests are that desperate for a drink, they can bring a flask and go hide out in a bathroom stall. Not your problem! Then again, maybe you are generous-hearted, care more about everybody having a good time than I do, and really don't mind being around alcohol. By all means, go ahead and provide drinks. But please offer a delicious signature mocktail as well.

You could even hire a special sober bartender. Chris Marshall, owner of Austin, Texas' Sans Bar, is probably the closest thing there is to a celebrity sober bartender. I got to meet him when he did a special popup bar in Portland, Oregon, and I've interviewed him over the phone several times. He's been hired for fun events like bartending for nuptials in the Hamptons, or for a two-day Indian American wedding. At the Hamptons event, he served non-alcoholic drinks for a wedding pre-party.

Then at the reception, he had a sober bar set up beside a bar serving alcohol. "And it was very, very cool to see how when you offer people an option, yeah, some people chose alcohol because they were going to go to the wedding to get very, very drunk and dance and do the whole thing. But some people were like, 'Wow, I'm glad this is here because I really did not want to do that,'" Chris said. He also noticed people regulating their drinking throughout the night by alternating cocktails and mocktails. "It was very eye-opening to me, because I just figured that everyone drinks like me. Which is like, more."

The challenges of bachelor and bachelorette parties

Uh oh. A bachelorette/bachelor party in sobriety? How arc we going to deal with that? These are traditionally some of the least sober events around. Have you ever seen a group of sober women wearing inflatable penis hats on their heads in public? Unlikely.

Bachelor and bachelorette parties top my list of events that it's okay for people in recovery to skip. But if you feel very strong in your sobriety, or if you don't drink for other reasons and only fear boredom or embarrassment, you might choose to attend.

Chris Marshall told me about attending a bachelor party in New Orleans with a bunch of his old college friends. Things were clear upfront: Chris was sober, they were going to get drunk. In fact, they started drinking as soon as their van crossed the state line into Louisiana. "They woke up drunk and they just kept at it," Chris said. "They were a bunch of dudes just getting wasted." Unperturbed, he spent time with his drunken friends but also carved out time for himself. "I was up every single morning and strolled down to the French Quarter. Had beignets for breakfast, and Café du Monde coffee. I just made sure that I was really treating myself to some time, walking along the Mississippi as the sun was coming up. Those are memories that I have." His friends remember little at all.

Interestingly, his friends looked out for Chris. "All those guys, as drunk as they got, they never once pressured me to drink alcohol. In fact, they went out of their way to make sure that I had a Coke, or I had a tonic with lime. And I felt very, very loved and cared for." For Chris, being with his friends was worth it.

But such a trip is not for everybody. "Check yourself," he said. "If people are drinking and you know they're going to be boozing it up pretty hard, make sure you're in a good headspace to deal with that." Chris also had an exit strategy. "We all came in that van. But I knew that in an hour I could get a flight back home if it ever got too much for me." He also brought his favorite non-alcoholic drinks.

Although she was an experienced sober traveler with many years of recovery, Sarah O'Brien had a different experience as the only sober person at a bachelorette party in Nashville. "It wasn't the thought that I wanted to drink at that moment," she said, "but I was very uncomfortable with the culture as a whole. You're in the middle of Broadway, there's 15,000 people, every bar is jam-packed from floor to ceiling." She felt triggered by seeing so many drunk young women wandering around. "It's hard to watch knowing that that's how I used to act. I had zero care in the world. And I always took it to the utmost extreme. But seeing it with a clear head, it was extremely overwhelming." Now that she has a daughter approaching her teen years, it's even harder to watch. "I was certainly fine. But at that moment,

I kind of realized, maybe Nashville isn't the place for me for a bachelorette party." She went home early.

Sober traveler Rebecca Rush reminds us that opting out is a viable choice. "Send a gift if appropriate and stay home. Nobody really wants a sober person at these events anyway, don't kid yourself. I don't care if it's your stepsister's bachelorette, you do not have to go."

Whether it's the bachelor party or the wedding itself, before automatically RSVPing yes, ask yourself: Do I have to go to this? Do I want to go to this? Would it be better for me not to attend? If in doubt, "Earlier on in your journey, my recommendation is just don't go," sober coach Amanda Kuda said. As she tells her sober coaching clients, "You really overestimate how little your presence is necessary at a lot of things. If it's going to compromise your mental wellbeing, it's going to compromise a commitment you've made to yourself, just don't go. People will get over it. We have this sense of almost like self-importance that our presence is going to make or break something. And as lovely as we each are as individuals, this is not true."

When you're the party planner

Planning a bachelorette or bachelor party for a sober person? Instead of the usual drunken hangout, plan around the interests of the bride- or groom-to-be. Do something outdoorsy, like kayaking; or culinary, like cheese tasting; or competitive, like a board game night; or relaxing, like going to a Nordic spa. Make sure that guests know upfront it's a non-alcoholic event, and that they agree to refrain from drinking for the duration of the party.

SOMBER AND SOBER AT FAMILY FUNERALS

Funerals are another common relapse trigger. Here's why:

- Grief, obviously

- Fear—OMG, this is going to happen to me someday! And to everyone I love!

- Social anxiety because you don't know how to act at a funeral

- Deep-seated resentments against family members or old friends present

- Secret celebration because you hated the person

- Discomfort because you didn't know the person all that well and you're surrounded by people crying and carrying on

- General feeling of creepiness in the presence of a dead body

I'm sure I left out many, but you get the idea. Lots of people don't like dealing with feelings, and these kinds of events stir up the biggest emotions. Throw in family and appearances by friends, enemies, and frenemies from your past, and suddenly you might find yourself very, very thirsty.

How to protect yourself? Same as in other events: get your own hotel room, organize a support system or enlist an ally, give yourself permission to leave early. Research transportation ahead of time. Is there Uber or other rideshare? Taxi? Bus? Have the numbers, have the apps, look at the bus route. Even if you don't use any of those things, knowing an escape route will help you feel less trapped.

FREAKING OUT AT MY GRANDMOTHER'S WAKE

My delightful grandmother died when I was three years sober. She was the only grandparent I knew—the rest died before I was born—and I grew up close to her. She came over from Ireland when she was a very young woman in the 1920s and never lost her accent. She had a small network of Irish friends who would pop by to visit, sing, and play the accordion. One of my fondest memories is of my sisters and me dressing up in her nightgowns pretending to be cocktail waitresses. We'd refill the whiskey glasses of our grandmother and her friends, and they'd tip us a dime or quarter.

I flew home for her funeral. Afterwards, we had a small family gathering in her house. I'd spent so much time there with her and had so many memories. There was nothing now but grief, a handful of relatives—some of whom were on the brink of fighting over inheritance—a platter of brownies and a table full of bottles. Not surprisingly, I got the urge to grab one of those bottles and empty it.

I snuck off to the phone—this was way before cell phones—and called a local person I knew who had been one of my mentors in early sobriety. Fortunately, she picked up. Unfortunately, she told me in a chipper voice that she'd realized drinking hadn't been her problem after all, so she'd returned to it. This was about negative 100 percent helpful. I hung up the phone, my heart racing.

My parents and sisters knew I'd quit drinking. But it wasn't something we talked about a lot. Alcoholism still has stigma as I write this, but even more so then—especially for a young woman from a nice middle-class family. I was embarrassed. They were embarrassed. But now I found myself desperate.

I went outside on my grandmother's porch to quietly freak out. My dad followed me. We'd had a fraught relationship all through my teens, mostly because by the time I was 14 I figured I was grown up and could do what I wanted and he was always getting in my way. It wasn't a confidante sort of relationship. But in my desperation now, I told him how much I wanted to drink. His response? "We're going. Now."

He stormed back inside and announced that my immediate family—about half of the small gathering—was leaving that second. I trailed behind him, saying, "No, we can't." I mean, the party was in my grandmother's honor. But at that second, my desire to drink was lifted. I had never understood how much my dad had my back. He looked totally demented barking at everybody that they had to leave and offering no explanation. I was stunned that he cared so much about my wellbeing that he didn't spare a single thought that to everybody else, his inexplicable behavior made him look like an asshole.

This was a huge turning point for me, both in my relationship with my father and in realizing that there are potential allies in lots of situations, whether or not they're sober themselves. It also gave me an important insight into Dad. How many times had his brusqueness been protecting somebody else, and I just thought he was a jerk? Understanding the depth of his support made it possible for me to calm down and be there for the rest of my grandmother's wake.

SPECIAL CONCERNS FOR SOBER TRAVELERS OF VARIOUS IDENTITIES

Sober travelers come in many flavors—and identity can intersect with sobriety in surprising ways. Race, age, ability, religion, and sexual identity can all influence people's experience of sobriety. Here are a few examples. I'm sure there are many more.

A SOBER VETERAN'S STORY

Phillip Vitela was 23 when he joined the military. He didn't realize he would make a career of it. But then 9/11 happened, and he stayed in until 2016. He grew up in military culture—the good and bad parts.

"The military is the first to integrate everything," he said. "You really rarely experience any kind of racism in the military." But drinking was even more prevalent than in civilian culture, an attitude that's just starting to change. "The more you could drink, the cooler you were. Every event we had was full of alcohol. Every weekend. I don't know how we survived."

Later in his career, Phillip worked internationally in special assignments. "So I went to a lot of places where you wore civilian clothes and you met diplomats, or you were out in town meeting foreign contacts or whatever. Wherever you were at, you drank with those guys. That's how you break the ice. You drink and smoke horrible cigarettes. And you're just—you're in it. And you're depressed because you're gone all the time. You're bonding by getting drunk with your fellow soldiers and the people you're working with. And there's no controls. There's no right or left limits. Nothing. It's actually kind of encouraged. As a matter of fact, when you're meeting a foreign diplomat, you're going to drink. Whether you like it or not, it's happening."

Both during and after his military career, he's encountered people who make a lot of assumptions about him based on his service. In Southeast Asia, he and a soldier friend walked into a European bar. "Immediately a girl came up. She said, 'Oh, the Americans are here. You kill anybody today?'" People have also assumed that he's a Republican, homophobic, and loves to hunt.

Now retired from the military and in his early fifties, Phillip sees how hard it is for his fellow veterans to adapt and assimilate. And to cut back on their drinking. "You grow up in the military. You travel, you meet a million people. But if you went in when you were 19 and you grew up in the army, the ages vary so much. So you're still hanging out with people who are 19, 20, 30, whatever it is. You get out and you still think you're in your twenties, but you're not. All of a sudden, you're alone." And when you travel somewhere as a civilian, it can be a very lonely and disorienting experience. "So traveling for veterans, and I've talked about this with my friends, also—we don't like to travel. I am so much more comfortable in my house. It's difficult for me to get out there."

Now that he's a couple of years sober, he's starting to venture out more, including discovering the world of going out for mocktails. "Being a veteran and traveling, you just have to forget you are a veteran. Get over that part, because that's the past. You have to leave that behind you."

AFRICAN AMERICANS AND ALCOHOL

While African Americans drink less alcohol than whites—and many abstain completely—they're likelier to suffer alcohol-related health problems. For example, death from cirrhosis is 1.27 times more common for Black drinkers, and they face a 10 percent higher death rate from alcohol overuse. Other common consequences from drinking are oral cancer, hypertension, stroke, liver disease, and depression.

"We theorize that this effect is due to the complex interaction of residential discrimination, racism, age of drinking, and lack of available standard life reinforcers (e.g., stable employment and financial stability)," said researchers in a Johns Hopkins study.[15] African American youth are also exposed to more alcohol advertisements than other ethnic groups.[16]

Fortunately, Black entrepreneurs are forging new paths. Pauline Idogho, founder and CEO of Mocktail Club, is currently the treasurer of the Adult Non-Alcoholic Beverage Association. "It hasn't been there for hundreds of years like the alcohol world so it's great for us to step in and help shape this world and this

---

15 "Less Drinking, Yet More Problems: Understanding African American Drinking and Related Problems," www.ncbi.nlm.nih.gov/pmc/articles/PMC3758406

16 "African-American Youth Exposed to More Magazine and Television Alcohol Advertising than Youth in General," publichealth.jhu.edu/2012/jernigan-african-american-alcohol

new industry," she said of Black business leaders in an article published on the Travel Noire website.[17]

SOBER LGBTQ+ TRAVELERS

A study by the Substance Abuse and Mental Health Services Administration (SAMHSA) released in 2023 looked at use of alcohol and other drugs in gay, lesbian, and bisexual people.[18] The results showed:

- Lesbian and bisexual females were more likely than straight females to have engaged in binge drinking in the past month, and about twice as likely to have engaged in heavy drinking in the past month.

- Gay and bisexual males and females were two to three times more likely than their straight counterparts to have used illicit drugs other than marijuana in the past year.

- About one third of bisexual females, bisexual males, and gay males had a substance use disorder (SUD) in the past year. About one fourth of lesbian females had an SUD in the past year.
  And as many of us know, a SUD loves to hop in our suitcase and come along for our travels.

---

17 "Black Women are Revolutionizing the Non-Alcoholic Travel Movement," travelnoire.com/mocktail-club-non-alcoholic-beverage
18 Unfortunately, this study is no longer available online. It was one of the many victims of radical right purging in early 2025.

For Vancouver, British Columbia resident Tamar Medford, being queer adds another level to travel and to sobriety. "I have the genetics for addiction and alcoholism," she said. "But I think part of the reason I started drinking was because I knew I was different. And I didn't want to recognize that."

Her partner is more comfortable being out. "My partner's like, 'Oh, you should get this little rainbow thing for your jeep.' And I'm like, 'I don't want to put anything rainbow on my jeep.'" Tamar doesn't want to draw attention to herself, especially when she's driving through some parts of the United States. "Even though we shouldn't have to hide who we are, there still is that fear of having to watch your back and making sure you go to places that are LGBTQ friendly." She loves to meet people but finds herself watching what she shares, especially if the other person seems religious. "It's not safe everywhere," she said. "It doesn't feel good. I've heard that Vancouver is actually the most queer-friendly city. But even here, there's a lot of communities that are incredibly religious. So we always have to be mindful."

Being sober helps her be more aware during her interactions. "It puts you in a better position because you can read people better. If I have a conversation with someone, and they start talking about God, for example, I love having a conversation about God. But I will generally be mindful of not sharing that side of myself. And if you were drunk, well, all that goes out the window because you wouldn't really care. Being sober keeps you safer in so many ways."

FIND A CULINARY QUEST

We don't have to miss out on the world of tasting stuff because we don't drink. Even if we skip the wineries, there are lots of other things to drink or eat.

"Especially if I were new to not drinking, I would definitely really dive into food and tea culture," said Sans Bar founder Chris Marshall. "Because it offers many of the same kind of trappings as alcohol that I think we kind of like as adults. We want to know history, culture, process, sourcing. Understanding that tea has just as much history and culture and nuance as wine is something that's cool."

Many people enjoy having a quest when traveling. Personally, I always make it a point to try local vegan ice cream, if available. Maybe you gravitate towards pretzels or cheese or kombucha or tacos. And just think, next time your sister-in-law bores you about merlot or pinot noir or whatever (personally, I could tell you more about Mad Dog and Thunderbird) over a Thanksgiving meal, imagine dazzling them with your search for the ultimate pickle. Please be sure to drone on just as long as she did.

While researching this book, I heard from people touting great culinary quests around the world. Nondrinker Sam

Bellantoni recommends Korea's street food and tea. "I'm constantly looking for tea shops all throughout Seoul," he told me. "It's always fun for me to try out different tea places, from contemporary to more traditional. Through tea, I frequently come to understand a great deal about the local culture."

Many places have a coffee scene, from coffee houses in Budapest that stay open until midnight, to surprising finds like the Ozarks Coffee Trail in Springfield, Missouri. If your teeth can take this much sugar, the Sprecher Brewing Company in Glendale, Wisconsin is perhaps as well known for its sodas as for its beer. Its root beer and cherry cola have won awards. Stop by for a tour and try unlimited samples of their craft sodas. "We also hold mocktail events in our taprooms and regularly publish NA mocktail recipes on our website," Tim Cigelske, Sprecher director of communications, told me.

One cautionary note: sometimes alcohol lurks in food, and sober travelers have varying amounts of sensitivity to it. Some of us avoid kombucha, which has a very small amount of alcohol. Others avoid foods containing any alcohol whatsoever.

"The biggest issue over the past 20 years has been fending off food with booze in it," said Eric Lucas. Eric and I co-founded the Society of American Travel Writers' Sober Travel Affinity Group. "Many people think this is silly; if they press, I point out their ignorance. If I were recovering from heroin addiction, would they deride me for not wanting heroin in my chowder? And of course, many people, including chefs who should know better, mistakenly believe cooking removes alcohol, which it does not." While it's true that some alcohol burns off the longer you cook it, much can remain. An alcohol burn-off chart published by the USDA shows that after baking or simmering a dish for 15 minutes, 40 percent of the alcohol remains. Alcohol-flamed dishes—such as saganaki, the dramatic Greek dish that involves dousing cheese with brandy and lighting it on fire— retain 75 percent of the alcohol.[19]

Desserts are especially alcohol-laden. Tiramisu can be potent. Rum cake and bourbon-filled bon bons you can probably see coming—but berries served in a dish of vodka? I got a mouthful of this once at my uncle's house, much to my horror. And come on, vodka is not dessert!

---

19 "How Much Alcohol Is Left In Foods When Cooking," blogs.extension. iastate.edu/answerline/2016/04/14/how-much-alcohol-is-left-in-foods-when-cooking

Sober trip idea: Isla Mujeres, Mexico

Tamar Medford likes to take the short ferry ride from Cancun to Isla Mujeres, which translates to Island of Women. "You can rent a golf cart and you can drive around the whole island. The views are just breathtaking. And then we will usually go to the main area, Playa Norte. It's got white sandy beaches with turquoise blue water. They have yachts out in the water and all these restaurants on the beachfront. You can get a meal for a very reasonable price. You just sit on a little lawn chair, and go in the water, and order food and drinks." Also in the Cancun region, Tamar recommends deep dish pizza at Playa del Carmen and swimming with turtles at Akumal.

SO YOU'RE GOING TO WINE COUNTRY

*E*very time I go to a travel event, PR folks are hyping yet another wine region. It seems that grapes are as ubiquitous as weeds. If you travel much, you are sure to encounter a wine region sooner or later. Fortunately, they're usually gorgeous places. And if you can peel your mind away from the wineries, areas billing themselves as "wine country" generally feature both outdoors fun and nice extras. As Amy Snook, who sometimes chooses to abstain from alcohol while traveling, told the *New York Times*, "These regions offer all the perks of luxury travel—wellness amenities, beautiful landscapes and fine dining."[20]

If fine dining is your thing, you'll find chefs in places like Sonoma and Napa Valley, California who pride themselves on using the best ingredients. These areas are increasingly offering non-alcoholic pairing options for multi-course meals, as well as the expected wines. Meadowcroft Wines in Sonoma Valley has an unusual zero-proof experience. "Guests receive a flight of four mocktails, inviting them to engage in the same olfactory and gustatory exploration as their wine-drinking counterparts," said David Messerli, Meadowcroft's director of marketing and

---

20 "Sober Travelers Find Something to Savor in Wine Country," www. nytimes.com/2024/04/17/travel/sober-travel-mendoza-tuscany-sonoma. html?unlocked_article_code=1.lE0.cSGB.FDLgFuo2xxwy&smid=url-share

strategy. "They're encouraged to discern aromas, flavors, and nuances, much like they would with wine."

On a group press trip to Healdsburg, California, I went solo while the other journalists tasted wines. I cruised back-country roads on a rental bike, sampled teas at a tea lounge and visited a delightful rose farm. I must not be the only non-imbiber who turns up there; since my visit, Healdsburg has created a mocktail trail with eight participating restaurants.

Tuscany, Italy is pushing olive oil tasting as an alternative for its sober visitors. Despite its name, Cavas Wine Lodge in Argentina is seeing sober visitors who choose yoga, hiking excursions, and spa treatments rather than throwing back a few glasses of wine.

Nondrinker Ernest White II enjoyed visiting wineries in northern Virginia with his cousins during a family reunion. "I absolutely thought that I was going to be bored to death," he said. "But actually, we had a great time. We had great conversations. And the food in these places was good. I was also always ready to be tapped for driving if necessary."

However, keep in mind that visiting wineries is an advanced gig for people in recovery. If you don't drink for other reasons, your biggest worries will be getting bored or disgusted by other people's behavior. But if you're struggling with sobriety, wine country can be slippery, so I'd choose a different vacation. Those who are both secure in their recovery and saintly can be designated drivers. But if you're like me and your hackles rise and resentments multiply at the thought of sitting around wineries waiting to chauffeur merry drinkers, just say no to that role.

Sober trip idea: A foodie vacation in Franschhoek, South Africa

Franschhoek is known as South Africa's food and wine capital. Sober traveler Eric Lucas and his wife Nicole didn't miss the alcohol at all during their visit. "We loved Franschhoek and easily ignored the wine country thing," Lucas said. He is determined not to eat any food cooked with alcohol, let alone drink it, and Nicole has a couple of food allergies of her own. "The dining establishments were superbly professional about that," he said.

The place's magic starts with its setting. "The valley is tucked in between forested 2,000-foot mountains and honestly put me in mind of Shangri-La. The foothills are vineyards and orchards dotted by stucco farmhouses and estates. Even in summer, the weather is cooler than just an hour away in Cape Town—75-85 degrees." He recommends Epice, a restaurant that focuses on spices. "Every dish prominently features one or more African/Asian spices, and they begin the meal by bringing you a spice cabinet holding the two or three dozen spices you'll have that night." They especially liked lingering over a sensationally atmospheric dinner on the open-air patio of an estate house overlooking the whole valley.

He and Nicole enjoyed shopping for handcrafted items, and the area also has easy access to penguins. "The best place to see penguins is two hours away, in Betty's Bay. Much less crowded than the penguin sanctuary just north of Cape Town. The penguins are as gosh-darn endearing as you expect, the setting is great, and you can't do that in Napa!"

SOBER BARS

round the world, alcohol-free bars have been opening as adult venues for socializing. Some have permanent brick-and-mortar locations, while others operate as pop-ups. From Bendición in Chicago to Hekate in New York to Club Soda in London, these places have wide appeal. They attract people in recovery, young adults 18-20 who aren't old enough to get into a regular bar, pregnant folks, first dates who want to get to know each other without alcohol, and old friends who enjoy catching up without yelling to be heard over loud music and bar noise.

Sans Bar in Austin, Texas is a forerunner in this area. Founder Chris Marshall was an alcohol and drug counselor who loved working with his clients. But all the counseling in the world didn't help when it came to their social lives tanking. "The help needed was not between the hours of eight to five," Chris told me. "The hours that people needed help were on a Friday night, on a Saturday night when they are all alone." When a series of his clients died alcohol-related deaths in 2017, he decided to try a different approach to helping them.

He opened Sans Bar in 2017. "The response at first was tepid, to put it nicely. There were a lot of people who just did not understand what I was trying to do. We got a lot of pushback from the public. Someone said, 'What's next, a restaurant that doesn't serve food?' I can laugh at it now, but back then, it hurt.

Those things cut really deep. And I learned not to read the comments on news stories. People were brutal."

Interest grew slowly. Chris' biggest surprise was that his top customers aren't in recovery. "Seventy-five percent of the people that come to Sans Bar more than twice a month identify as being either sober-curious or sober sometimes," he said. Early on, he even got hassled by some 12-step folks who worried that sitting in a bar and drinking non-alc drinks would be a slippery slope down to regular bars. Which is understandable. But Chris thinks recovery programs shouldn't define a person's whole identity. "I didn't get sober to live half a life," he said. He's striven to make Sans Bar a third place—besides work and home—that doesn't center around alcohol. "To me, Sans Bar is just this comfortable space where people can connect to each other and have drinks. The drinks are almost secondary to the community and connection that I want to create."

Sans Bar has been very influential. Chris founded Sans Bar Academy, a ten-week online course designed to help other people open their own sober bars. "I hope that in the next five years, we see 20, 30 other iterations of alcohol-free spaces and destinations across the country and across the globe," he said. Indeed, some of his graduates have already opened up their own places.

In Portland, Oregon, Andy McMillan's Suckerpunch operated as a pop-up sober bar, before he turned his attention to operating Heck, a non-alcoholic brewery. Andy was especially concerned with making Suckerpunch welcoming for people in recovery. "There's a particular dance there with creating a space that feels familiar and feels like a bar and has that kind of energy, and also understanding that a lot of stuff about bar culture or drinking could be challenging or triggering for folks who are in recovery," he said. "So we are constantly doing the dance of trying to figure out how to make it safe and comfortable and welcoming." Instead of super-sweet Shirley Temples, sober bartenders create unusual drinks with some bite. Andy favored ingredients that are unfamiliar to many people, such as aquafaba, which gives drinks a frothy texture; alcohol-free botanical spirits; and verjus, a non-alcoholic byproduct of winemaking.

In San Antonio, brothers Roger and Mike Sanchez opened HASH (for Heal and Spread Healing) Vegan Eats as a restaurant and sober bar. The space has the look and feel of a clubhouse, and serves San Antonio's less economically advantaged South Side. A photo of their late mother, Cynthia Ann Medrano, hangs over the door. She died of cirrhosis when her sons were young. "I remember being really annoyingly loud about my lifestyle, and like drinking and using drugs," Roger told me as we sat at a table in HASH. "And then when I got clean, what happens a lot in recovery is people want to be real reserved and stuff. Which is fine, but like, the person who I am is a very loud and outgoing person. So in recovery, I wasn't going to change that. Recovery saved my life. I mean, so why am I going to be shy about it?" I loved how these guys were living their message. They regularly host free stores. On Taco Tuesdays, they serve dollar tacos to neighborhood families.

Sober bar owners often get creative with events planning, so there's more to do than just sit and sip. In New York City, Hekate Café & Elixir Lounge books sober queer nights, readings by sober authors, astrology and energy healing sessions, and other events to attract new customers and keep regulars engaged—all in an atmosphere with a delightfully witchy and magical theme. They let me do a book signing there once.

Hekate has an interesting backstory. Owner Abby Ehrmann opened a regular bar called Lucky in the East Village in 2016, which she designed to be a welcoming space. "It was so good that people were coming every day and enjoying what I was creating. And then they were drinking every day," she told me on a Wednesday afternoon between refreshing her regulars' drinks at Lucky. "I was minting fresh new alcoholics."

Abby's top two customers helped inspire Hekate. "Customer number one had a stroke, and his doctor said, 'Maybe you shouldn't drink so much.' And customer number two got sober after the pandemic when we were all reevaluating our relationship with alcohol. Because you couldn't go to your local bar and have a glass of wine, but you could get a whole case delivered to your apartment. And people were doing that. And then they're like, 'What the hell, I just drank a whole case of wine!'"

Concerned that her two customers were having problems with alcohol, Abby asked herself what she could do to create that same kind of community she had at Lucky, but without the alcohol. Hekate was born in 2022.

WHAT ARE THESE NON-ALCOHOLIC BEERS, WINES, AND SPIRITS?

Things have gotten much more complicated and interesting in the non-alcoholic drinks sphere lately, especially in bigger cities. Instead of settling for a club soda with lime, you might see a menu that includes craft zero-proof cocktails and non-alcoholic beer and wine. If you're like me, at first you might feel confused and wary. Some of these ingredients sound suspiciously like booze. What is this stuff?

As "soberpreneur" Andy McMillan explained, "The spirit alternatives as they exist at the moment fall broadly into two categories. There's the knockoffs, the parities of whiskey or gin or rum or whatever. And then there's sort of more esoteric

spirit alternatives that don't have any context in alcohol at all." For example, the Ransom Farm & Distillery in Oregon came out with a line of NA spirits called Dhos. Its bittersweet spirit blends bitter rhubarb, bitter herbs, cinchona and gentian root with fruity stuff like watermelon, kumquat, yellow cherries, and baked oranges. It's sweetened with monk fruit and has only five calories per ounce.

In Hood River, Oregon, you can visit Wilderton, the first non-alcoholic distillery and tasting room in the U.S. Chief distiller Seth O'Malley comes from a tea-making background, so he knows which botanicals respond best to being distilled in water rather than alcohol. "We decided early on that we didn't want to make just, like, a non-alcoholic version of whiskey, for instance," Seth told me. "I wanted to use the non-alcoholic concept as an opportunity to create new spaces and new flavors and new spirits that weren't just trying to taste exactly like other types of spirits." He experimented with his collection of botanicals to develop complex, intense flavors. Currently Wilderton is producing two types of aperitivo, citrus and bittersweet. I've visited the tasting room and tried the aperitivos. While I was concerned about venturing into the world of non-alcoholic spirits, I have not felt triggered to move on to the real stuff. Of course, this is a personal decision.

When it comes to making non-alcoholic wines, there are several methods that can be used. They start with real wine, which is then dealcoholized using methods such as vacuum distillation or reverse osmosis. How do dealcoholized wines taste? Ariel Dunitz-Johnson shared their opinion with me as we sat outside at a Portland café. At press time, Ariel is trying to crowdfund a permanent home for their popup non-alcoholic bottle shop Ever After. They know a lot about fake booze. The dealcoholization process is hard on grapes, Ariel told me. "It kind of strips it down, waters it down, takes the taste out." Quality varies. "I'm pretty picky about NA wines." Ariel is partial to Oregon-made NA wine from Varnum Vintners.

Non-alcoholic beer producers may also use vacuum distillation or reverse osmosis, or they might arrest the fermentation process or limit fermentation with special types of yeast.[21]

ARE NON-ALCOHOLIC DRINKS FOR YOU?

While many people enjoy non-alcoholic beer, wine, and spirits, some people in recovery avoid them. People might be uncomfortable drinking something so reminiscent of their old frenemies.

Also, non-alcoholic doesn't necessarily mean alcohol-free. The FDA's definition of non-alcoholic is containing less than .5 percent alcohol by volume (ABV). "FDA does not consider the terms 'non-alcoholic' and 'alcohol-free' to be synonymous. The term 'alcohol-free' may be used only when the product contains no detectable alcohol," the FDA's website explains.

The alcohol content of a .5 percent ABV drink is very little when you consider that the average beer is 5 percent ABV. You'd have to drink 10 non-alc beers to equal a single regular one, and you would probably be too busy running to the bathroom to ever get drunk. Still, as sober writer Katie MacBride wrote in a Healthline article, "as someone who was so severely addicted to alcohol that some mornings I drank cough syrup or mouthwash just to get my hands to stop shaking, I don't mess around with even small amounts of alcohol." Fair point!

---

21 "Approaches to Non-Alcoholic Beer Fermentation" escarpmentlabs.com/blogs/resources/approaches-to-non-alcoholic-beer-fermentation accessed June 22, 2024

Then again, fruits naturally ferment as they ripen; a banana can contain .2 to .4 percent alcohol. The FDA site goes on to explain, "Beverages such as soft drinks, fruit juices, and certain other flavored beverages which are traditionally perceived by consumers to be 'non-alcoholic' could actually contain traces of alcohol (less than 0.5 percent alcohol by volume) derived from the use of flavoring extracts or from natural fermentation. FDA also considers beverages containing such trace amounts of alcohol to be 'non-alcoholic.' We, therefore, have no basis for objecting to claims of 'non-alcoholic' on labels of dealcoholized wines, even though they are derived from alcoholic beverages."[22]

There are also now truly zero-alcohol beers, such as Budweiser Prohibition Brew. If in doubt, choose the alcohol-free label over non-alcoholic, or skip this drink category entirely.

Sober traveler Eric Lucas said he's not drinking any beer which "supposedly had most alcohol sucked out of it. That's not okay with me. People sometimes deride me when I bring it up. 'But, Eric, it's such a tiny amount,' a neighbor of mine who's a scientist once snarled at me. Would anyone say this if we were discussing a so-called tiny amount of fentanyl?" Eric instead sticks with soda fountain drinks, thank you very much. "I'm uninterested in the question of whether a tiny amount would 'trigger' me or affect me or whatnot," he said. "I have committed to myself and the universe that I do not put alcohol in my body—not two gallons, and not two 'tiny' drops."

But for many nondrinkers who avoid alcohol for reasons of health, wanting to have a clear mind, or just taking a break, this category of drinks really expands their drink options, especially in social settings. And I love anything that normalizes

22 "CPG Sec 510.400 Dealcoholized Wine and Malt Beverages - Labeling" www.fda.gov/regulatory-information/search-fda-guidance-documents/cpg-sec-510400-dealcoholized-wine-and-malt-beverages-labeling, accessed 4/12/2025

nondrinking. I think these also encourage some drinkers to slow down. Given a choice, some people will alternate a regular drink with a non-alc drink to prevent inebriation. For example, Guinness Zero is a huge hit in Ireland. At press time, the 0 percent stout accounted for almost 5 percent of the Guinness brewed, and the company was dramatically increasing the availability of Guinness Zero on draft to pubs across Ireland. This bodes well for people's health, and for lessening automobile and other alcohol-related accidents.

MOCKTAILS IN REGULAR BARS

Again, hanging out in bars isn't the best idea for many folks in recovery. Makina Labrecque, regional bar manager for Concorde Entertainment Group in Calgary, advises, "Number one, don't do it if you don't feel comfortable. Knowing your own personal boundaries is very important."

But if you're one of those recovering people who doesn't feel threatened or tempted in bars, or if you don't drink for other reasons, you might find yourself in a bar looking for an NA drink. After Makina quit drinking, she took a break from working in her beloved hospitality industry and didn't go out much for the first year or so. But once she grew more comfortable, she found ways to navigate bars and restaurants. Before going to a new place, she looks at the bar menu to check out its selection of zero-proof drinks. If she finds it lacking, she's not shy about calling up and asking to bring her own bottle of NA wine. She's happy to pay a corkage fee.

"I see value in paying for that to have a good, inclusive experience. The reason people love dining out is that shared experience." She doesn't want to sit there with a Coke while everyone else has a flavorful cocktail. "Don't feel weird about asking. I haven't had anyone make a big deal out of it. I've always been told, like, yeah, it's totally fine."

As a writer who specializes in sober travel, I get an astounding amount of information about restaurants and bars around the U.S. and the world that are adding mocktail menus or at least a few alcohol-free specialties to their menus, especially around Dry January and Sober October. One of the most intriguing was that the Condado Vanderbilt, a historic hotel in Puerto Rico, has introduced a zero-proof rum alternative, which means that you can still sample the ingredients the region is known for—sans intoxication. I love to see this kind of normalization of not drinking alcohol. Smaller destinations like Pensacola and Saint Augustine, Florida and Knoxville, Tennessee have assembled "mocktail trails"—lists of local mocktail hotspots to appeal to the nondrinking public. Or, as Hood River, Oregon calls it, a Mocktrail. In Baltimore, sober bar manager Anna Welker has created an acclaimed zero-proof cocktail program at Hotel Revival.

Of course, bars are full of alcohol. There's always the chance the bartender will accidentally pour some in your glass. "I think it's kind of like allergies," Makina said. "As a diner, you have to assume the risk. I've always been a big fan of personal risk assessment. If you're feeling like the place doesn't care, maybe don't give them your money."

When ordering a non-alcoholic drink in a regular bar, you should enunciate, repeat yourself, and double and triple check. Bartenders may be busy, and the place can be loud. Don't be embarrassed to ask for reassurance that your drink is alcohol-free.

Chris Marshall recommended carrying drink recipe cards (which you can buy online or print out your own) in case you visit a bar that hasn't kept up with the non-alc movement. For example, an alternative to a mojito could be lime, muddled mint, simple syrup, and club soda.

After a few "virgin" drink fails, sober traveler and podcaster Tamar Medford avoids ordering replicant cocktails. Instead, she now orders smoothies. "Order something that you know does not have alcohol in it," she advised. If in doubt? "I would always recommend smelling it before you drink."

> "Los Angeles has plenty of restaurants with good mocktails and NA options. I think the West Coast in general is more accepting of the NA movement than NYC and the East Coast."
> —Jason, NYC-based publicist and sober traveler

Sober trip idea: Aruba

If you want to visit the Caribbean, sober traveler Sarah O'Brien recommends Aruba, which she's visited multiple times. "I'm a big beach person," she told me. "I like to spend 12 hours on the beach. Aruba is not as big of a party scene as some of the other Caribbean islands." The restaurants are great, too, and she appreciates the variety of people. "There's a lot of different age groups there. I think a lot of people that are retired and older, so not as much partying going on there. It's also a family destination. I have fun there with my girlfriends and cousins."

SPECIAL INTEREST SPOTS FOR NONDRINKERS

efraining from alcohol is nothing new. I find the historical fight against demon alcohol fascinating and am interested in sober history wherever I go. People have been struggling to quit and stay sober for hundreds of years in the U.S., and longer in other places around the world. While people often act like 12-step groups are revolutionary, they stand on the shoulders of previous mutual aid societies, such as the Washingtonians and the Oxford Group. Here are a few places that will interest nondrinking history aficionados. Many rely on volunteers and unstable funding sources, so check their current hours before visiting.

American Prohibition Museum

While overall the American Prohibition Museum in Savannah, Georgia has a pro-alcohol vibe, it also owns a large and fascinating collection of temperance memorabilia. Associate manager Juliana Sims told me: "In our gallery dedicated to temperance, we have everything from pins from the Women's Christian Temperance Union (WCTU), to pledges men signed to swear off alcohol, sheet music of Prohibition songs, as well as memorabilia that Carry A. Nation sold to fund her bar bashing mission."

Plus, there's stuff belonging to drinkers. "We have a case of all the different kinds of flasks that people made to hide their alcohol in plain sight, such as in eggs, books, and cameras." The museum tells the story of the fight for and against Prohibition with costumed docents, four vintage cars, more than 30 wax figures, and the recreation of a speakeasy. I loved the full-size display of Carry A. Nation pulverizing a bar with her famous hatchet.

Though they don't keep statistics, Sims was willing to speculate on the alcohol-use status of visitors. "We actually get a lot of sober guests here at the museum. As we always say, everyone out there has an opinion or relationship surrounding alcohol, whether it be good or bad. Our museum presents both sides of the argument surrounding Prohibition, so everyone who comes through, sober or not, can see their side represented and understand its counterpart. I would say around a third of our guests are sober, which puts us in the tens of thousands each year."

The strangest thing I saw at the museum was an old photograph of hooded KKK members riding a Ferris wheel. Unfortunately, as I learned, some factions of the temperance movement made an unholy alliance with the Klan, who rallied around Prohibition. These terrorists used Prohibition enforcement as an excuse to violently run bootleggers out of town, and go after the immigrants they thought were ruining the country with booze. Mostly these were Germans and Irish Catholics. (Hey, my forebears!) Despite the Klan involvement with some parts of the temperance movement, it's important to note that there were also Black and Native American temperance organizations and leaders; abolitionist Martin Delany organized the first Black temperance society in Pittsburgh in 1834.

The most disappointing thing about the Prohibition museum was the lack of temperance memorabilia in the gift shop. I had hoped to do all my Christmas shopping there. But alas, it was mostly shot glasses and that sort of thing.

Stockade Museum & Carry Nation Home, Medicine Lodge, Kansas

Speaking of Carry Nation, she is one of my favorite historical characters. Born Carolina Amelia Moore in 1846, she married a drunkard at a young age. He died from drinking and left Carry (sometimes her name was spelled Carrie, but Carry A. Nation worked better for her purposes) alone to raise a small child. She later married preacher/lawyer David Nation and they settled together in Kansas, where Nation got involved with the Women's Christian Temperance Union. In 1880 Kansas became the first U.S. state to ban manufacturing and selling alcohol. However, Carry didn't feel that the ban was being sufficiently enforced. She split with the nonviolent WCTU and began to take her hatchet to barrooms, smashing up saloons. Many people, then and now, regard her as a zealous kook. But she had good reason to hate alcohol. At a time when women had few rights and fewer avenues to earn a livelihood, having a drunken husband was way more detrimental than it is today. And she was so bad-ass—six feet tall, not conventionally beautiful in the least, wielding a hatchet. Oh, to hear all those bottles smash!

Carry lived in Medicine Lodge, Kansas from 1889 until the early 1900s. You can visit her former home and peruse her photos and memorabilia. You can also see an 1886 jail, a furnished log cabin, a collection of fossils, and other random stuff donated by townspeople.

Oregon State Hospital Museum of Mental Health, Salem, Oregon

Back in the late 1800s, you might have wound up in the Oregon State Hospital if you were mentally ill, intellectually disabled, a drunk, a prostitute, or if your husband was sick of you. So many reasons to be committed!

When the hospital was first built, it was a model of forward thinking developed along the Kirkbride Plan. Dr. Thomas Kirkbride (1809–1883) was a psychiatrist who advocated for a system of moral treatment based on compassion and respect. He believed in fresh air, attractive architecture, and more freedom than some rival treatment philosophies. But he had a pretty broad definition of insanity and guesstimated that 1 in 500 people were insane. The most usual reasons for people to enter 19th century institutions were grief, ill health, anxiety, and intemperance. Other causes included religious excitement, prolonged lactation, tobacco use, metaphysical speculation, nostalgia, and exposure to the sun's direct rays.[23]

The Oregon State Hospital opened in 1883. As the field of psychiatry evolved, the hospital adopted new therapy approaches as they came out. What looks cruel and unusual today was once on the cutting edge. It was actually a very impressive place where the inhabitants grew most of their food on a farm, learned to make furniture, prepared meals for up to 3,000 people per day, and even sewed their own straitjackets!

While there aren't any displays specifically on treatment for addiction, recovering alcoholics can get a feel for what their home might have been like a century ago. As museum volunteer Kylie told me, "There are references in the museum as a whole to individuals that had been diagnosed with dependencies on alcohol and other substances, but as part of a larger conversation about life and experiences at the hospital. For example, in our 'more than a statistic' section there are lists of demographics of hospital population based on reason for admission that lists drug and alcohol dependence."

The museum is in the original part of the hospital, adjacent to what's now the working hospital for forensically committed patients, aka the criminally insane. The museum is fascinating, full of arcane artifacts and even the real voices of patients whose stories were recorded; visitors can pick up vintage telephones to listen to first-person accounts about being admitted to the hospital.

Richardson Hotel, Buffalo, New York

You can spend the night in what was once the Buffalo State Asylum, reopened as a swanky hotel in 2023. Just think,

23 "Dr. Thomas Story Kirkbride," www.uphs.upenn.edu/paharc/time-line/1801/tline14.html accessed May 6, 2024

alcoholics could have been committed here in the late 1800s—and now, people in recovery (or functioning alkies who haven't lost all their money yet) can choose to rent a room for the night! The building was designed following the Kirkbride Plan, with grounds designed by Frederick Law Olmsted, who is considered the American father of landscape architecture. The huge imposing brick structure has National Historic Landmark status, 40 acres of green space, a local history museum, and a self-guided audio historic tour (plus modern amenities like Wi-Fi and fine dining). When I visited, I took an excellent hardhat tour that lets you see the wards that have not been upcycled into hotel rooms. Expect peeling pink and green paint, 100-year-old abandoned wheelchairs, and feelings of melancholy.

Temperance Hotel, Ladysmith, British Columbia

This historic lodging option is on Vancouver Island, about 55 miles north of Victoria. You can spend the night in the hotel, which was built in 1900 to house coal miners. In earlier days, Ladysmith had a large population of rowdy single men living in hotels above saloons and restaurants. The Temperance Hotel was the only lodging that didn't serve alcohol. It's also known for housing strikebreakers during the Great Strike of 1912-1914—for which it received two small bombs, one blowing off the hand of a strikebreaker as he tried to dismantle it. The Temperance Hotel was listed on the Canadian Register of Historic Places in 2015 and reopened as a boutique hotel in 2023, after a long restoration project.

Frances Willard House Museum and the Frances E. Willard Memorial Library and WCTU Archives, Evanston, Illinois

Author and activist Frances Willard lived in this house during the years she served as president of the Women's Christian Temperance Union. The house also served as an informal WCTU headquarters and housed some of the group's workers, and has been a museum since 1900. If you visit, you'll find many of the original furnishings used by Willard, including artwork, family photos, furniture, and even her bicycle. In 1965—100 years after the house was built—it was deemed a National Historic Landmark. People interested in the temperance movement or in historic houses in general will enjoy touring the Willard house.

Wilson House, East Dorset, Vermont

If you belong to or are interested in Alcoholics Anonymous or other 12-step programs, you might already know that AA was founded in 1935 by Bill Wilson and Bob Smith. But did you know that you can tour several of their former homes?

Wilson was born in East Dorset, Vermont, home of his paternal grandparents, and spent much of his childhood there. He continued to visit his grandparents into young adulthood. What's known as the Wilson House was constructed in 1852 and operated as a hotel for many years, through many changes of ownership. By the 1980s, the building was decrepit. Only its connection to Wilson saved it from demolition.

In 1987, a man named Albert "Ozzie" Lepper, an admirer of Bill W. and the AA movement, purchased the old hotel. He restored the property and reopened it in 1988 as the Wilson House. And yes, you can stay there! It's a pilgrimage site for AA members from around the world. Now a nonprofit foundation owns the hotel and volunteers run it. In keeping with AA principles, the hotel doesn't advertise. Guests are drawn through attraction rather than promotion. Even if you don't stay there, you can attend an AA meeting on the property.

Hardcore Bill W. fans can take a whole self-guided tour of spots that were important in Wilson's life. The Griffith Library, located on the Wilson House property, is the former home of Bill's maternal grandparents, and it now houses 12-step and spiritual recovery materials. You can see where Bill W. attended high school, the quarry where three generations of Wilson men worked, and the cemetery where Bill and his wife Lois now rest in peace, among other spots.

Stepping Stones, Katonah, New York

An hour north from New York City in Westchester County (and an easy day trip by train) is the home of Bill and Lois Wilson, founders of AA and Al-Anon, respectively. They moved into the house, which is called Stepping Stones, in 1941 and lived there until their deaths. In 2012 the National Park Service and the Department of the Interior designated the house a National Historic Landmark, and visitors can tour the house and its grounds.

Along the tour, a volunteer guide tells you about the love story between Lois—a rich girl from Brooklyn Heights who summered in Vermont—and Bill, a boy four years her junior who didn't come from money. Bill started drinking while serving in World War One. When he returned, he made big bucks as a stock analyst on Wall Street before the one-two punch of alcoholism and the 1929 stock market crash brought him to his knees. Lois stuck by him despite some incredible lows, including an inability to have the kids they dearly desired. Instead, they helped save a gazillion alcoholics.

People familiar with AA will find many things to ooh and ah over, including the kitchen table where Bill sat with a childhood friend who convinced him to try to get sober one last time; the original AA meeting coffee pot; the original "Came to Believe" painting (also known as "The Man on the Bed"); and the desk where Bill wrote *Twelve Steps and Twelve Traditions*. As Levi, a tour guide, told me as we walked slowly through Stepping Stones, "It's not a good tour unless somebody cries." On my tour, there were a few tears and a few chills running down our spines as we gazed upon sacred AA artifacts.

Dr. Bob's Home, Akron, Ohio

Also a national historical landmark for its central role in the establishment of AA, Dr. Bob's home fittingly has 12 steps to reach the door. It was built in 1915, and Dr. Bob and his wife Annie moved in the following year. They lived there until their deaths in 1949 and 1950.

Dr. Bob's Home is more hands-on than Stepping Stones. When I walked up the stairs and opened the front door, a volunteer greeted me with, "Welcome home!" He encouraged me to touch everything in the house, which is set up to reflect how it looked in the 1930s at the birth of AA. Visitors can see intriguing relics like Annie's 1930s era cigarette rolling machine (which her kids delighted in filling with pencil shavings to make her cigarettes catch on fire), the clothes chute where Dr. Bob once hid his bottles, and an inscribed walking stick Bill W. brought his friend from a trip to Ireland just months before Dr. Bob died.

Twelve-step members will be especially moved by standing in the rooms where the earliest AA meetings were held. Many early members stayed here as Dr. Bob and Annie—sometimes called the mother of AA—watched over them and tried to help them quit. I had a visceral reaction to a bottle of detox tonic sitting on a nightstand in the bedroom reserved for sobering up drunks. Dr. Bob mixed up this concoction of sauerkraut, Karo syrup, and tomato. As I tried not to heave, a volunteer explained to me that these folks "probably didn't have any vitamins in them and they didn't have any energy in them."

There's also a gift shop where you can buy historical literature, sobriety tokens, coffee cups, and other things relevant to AA members. While you're in Akron, don't miss the rest of the AA pilgrimage trail, including the longest-running AA meeting, the gatehouse of the Stan Hywet House where Dr. Bob's friend Henrietta Seiberling first introduced him to Bill Wilson, and the Smith gravesite.

Temperance fountains

Sure, you're highly unlikely to travel just to see a drinking fountain. But while you're out and about, you might see a fountain of historic interest to nondrinkers. In the late 1800s and very early 1900s, it was generally cheaper to buy beer than to get clean, safe drinking water. The 1874 WCTU meeting encouraged attendees to construct temperance fountains in their hometowns, so pro-temperance private benefactors paid to install public drinking fountains. Some are extremely grand, such as those built in New York, Boston, and Washington, DC. But you might find them in smaller, unexpected places as well. Shenandoah, Iowa has two especially ornate fountains installed by the Women's Christian Temperance Union in the early 1900s. One of them even made the National Register of Historic Places. I once stumbled across a WCTU drinking fountain in Anacortes, Washington. It always makes me feel a part of something to remember how long people have struggled against the problems caused or exacerbated by alcohol.

My city of Portland is especially abundant in temperance fountains, though most people probably don't know that's how the fountains came to be. Teetotaling lumber baron Simon Benson donated $10,000 to the city to construct 20 bronze drinking fountains around Portland. Known as Benson bubblers, you can still drink from them today—if you dare.

In the United Kingdom, temperance advocates supported the Metropolitan Drinking Fountain and Cattle Trough Association, which built drinking fountains across from public houses.

Smith's Cove Old Temperance Hall Museum, Nova Scotia, Canada

Built between 1834 and 1837, this is one of the oldest temperance halls in Nova Scotia. Baptist temperance advocate Benjamin Potter led meetings here, as well as Baptist services. Various temperance groups used the hall all the way up until 1972! This building now serves as a museum chronicling the area's history.

Marr Residence, Saskatchewan, Canada

If you find yourself in Saskatoon, Saskatchewan, stop by the Marr Residence. This National Historic Site was part of a temperance colony that predated the city. The utopian Temperance Colonization Society, formed in Ontario in about 1881, was part of a government scheme to settle western Canada and "recreate the best features of Anglo-Canadian civilization therein." It rounded up more than 3,000 settlers who wanted to be part of an alcohol-free community.

Things didn't go smoothly for the settlers. During the North-West Rebellion in 1885, wherein Metis, Cree, and Assiniboine people rose up against the Canadian government, the Marr Residence turned into a field hospital with eight doctors and six nurses. Since this was the first time in Canadian history that nurses cared for casualties, the Marr Residence is considered the birthplace of Canadian military nursing. Built in 1884, the Marr Residence is the oldest building in Saskatoon that remains on its original site. The residence is open to visitors sporadically, so check the current schedule.

Norsk Museum, Norway, Illinois

In the late 1870s, sick of newspaper reports of alcohol-driven stabbings, riots, rock throwing (an entry level weapon), and attempted murder, some Illinois women got together and established the Norway Temperance Association. In 1880, the 200-seat building for the Norway Red Ribbon Reform Club opened. The building was replaced in 1909, and it's been used for many things over the last century. Now it houses the Norsk Museum, which chronicles the area's Norwegian history inside the old temperance hall.

Keweenaw National Historical Park Calumet Visitor Center, Michigan

Way up in Michigan's Upper Peninsula, that piece of land that juts out from the rest of the state and is bordered by Lake Superior, Lake Huron, and Lake Michigan, there was once a Finnish temperance movement. Margareeta Johanna Konttra Niiranen (1861-1927), who went by the name Maggie Walz, was its charismatic leader. She arrived in the United States at the age of 20, learned English in night school, and worked her way up from a domestic servant through a series of jobs, including door-to-door saleswoman, notary public, newspaper editor, and founding member of a Finnish temperance society called Pohjantahti.

A strong proponent of women's rights, Walz helped other Finnish women immigrate to the United States, underwriting their passages, helping with job placement, and teaching them English. She established a Finnish colony on Drummond Island in Lake Huron based on Christianity, cooperative capitalism, and temperance.[24]

Modern day visitors can learn about Walz's life in the permanent exhibit at the Keweenaw National Historical Park Calumet Visitor Center. "Pine Street here in Calumet was prominent in the local Finnish-American business community," Jeremiah Mason, archivist for the Lake Superior Collection Management Center, told me in an email. Visitors can still see many of the buildings or sites where Maggie Walz lived and worked. "There are halls that were used by temperance and social groups that she belonged to that still stand as well."

You can also visit Drummond Island. The Drummond Island Historical Museum houses some of Walz's photos and writings. This would be a fun trip for outdoorsy nondrinkers, as the island is known for hiking, birding, kayaking, golf, an ATV off-road trail system, and shipwreck scuba diving.

Westerville History Museum, Ohio

The Anti-Saloon League was a major force in U.S. politics until 1933. Founded in 1893 in Oberlin, Ohio, it moved its headquarters to Westerville, Ohio in 1909. The league's presence in Westerville gave the town the nickname "Dry Capital of the World." The Anti-Saloon League's publishing arm, the American Issue Publishing Company, produced and mailed so much temperance literature that Westerville became the country's smallest town to have a first-class post office. The

---

24 "Maggie Walz," miwf.org/timeline/maggie-walz/ accessed May 22, 2024

Westerville History Museum features rotating exhibits on temperance and other parts of the town's history.

If you like browsing archives, this museum has one of the largest temperance and Prohibition collections in the world. I spent two whole days there as a birthday present to myself, poking around in the archive and taking a self-guided tour of the Temperance Row Historic District where the league's leaders built their houses.

Brown University Library, Providence, Rhode Island

For archive lovers, Brown University has the Alcohol, Temperance and Prohibition Collection of digitized publications from the Alcoholism and Addiction Studies Collections. You'll find pamphlets, government publications, and sheet music, so you can revive the old temperance songs.

The Mob Museum, Las Vegas, Nevada

The full name of this Vegas attraction is the National Museum of Organized Crime & Law Enforcement. Overall it celebrates drinking more than sobriety, but its Path to Prohibition exhibit is fascinating and includes temperance memorabilia. If you're feeling steady enough to resist the basement speakeasy, there's a lot of interesting stuff about the history of liquor consumption and distribution in America. Since it's the mob, prepare for violence. I found the museum well-done but quite gory.

Andrew J. Volstead House Museum, Granite Falls, Minnesota

This house museum commemorates the life of Andrew J. Volstead, son of Norwegian immigrant farmers, who served as a representative from Minnesota's 7th Congressional District from 1903-1923. It's been a National Historic Landmark since 1978. Volstead is most famous for authoring the Volstead Act, which prohibited the manufacture and sale of alcoholic beverages—commonly known as Prohibition. It remained in force from 1920-1933. Volstead was also behind the Capper-Volstead Act, which allowed farmers to join in cooperatives to bargain for better prices. His wife, Nellie Volstead, was a leading suffragist. The museum highlights their work. It's open on Saturdays. Sometimes it hosts fun events like escape rooms or Christmastime visits from Santa and real reindeer.

PICKING A DESTINATION FOR A SOBER VACATION

AVOIDING DANGER SPOTS

I firmly believe that if you have a good reason for being there, you can go anywhere without drinking. But sober travelers will notice that alcohol is more prevalent in some places than others. In fact, some cultures drink a lot. *A lot.* Then again, in some countries, alcohol is not in your face at every turn. I find this extremely relaxing.

If you're looking to avoid places with a lot of drinking, North Africa and the Middle East are top picks. Many of the countries in these regions have close to zero consumption.

On the other end of the scale, we have Europe. Prague tops the list of cities with most bars per capita, and Czechia has the highest alcohol intake in Europe at 15 liters per person per year. Belgium, France, Germany, Portugal, and Ireland are right behind, with 12 to 14 liters. In the U.S., Las Vegas, Orlando, San Francisco, and Miami all made the top ten list for most bars per capita. Of course, there's drinking and then there's binge drinking, which really gets people in trouble. In the U.S., for example, Wisconsin leads the list for highest beer consumption, with a 25.8 percent binge drinking prevalence.

These factoids aren't to discourage you from traveling wherever you want. You might find sober companionship anywhere. After all, Burning Man has sober camps, and some big music festivals, like Floydfest in Virginia, now have AA meetings. But in the spirit of "forewarned is forearmed," here are a few places my informants told me were especially hard to avoid alcohol.

- "Most restaurants I went to in Greece, at the end of the meal, they'd bring you a shot of whatever their specialty alcohol is without asking. They're very hospitable and kind, and they really want to share their culture. And that's a big part of their culture so they kind of pressure you to drink it, and it's hard to turn it away. Most of the time I would just accept it and then toss it or give it to someone at a different table. So that could be hard. And there is, especially on the islands, just a huge party scene. Like the sun goes down and it's a party. It is possible to avoid that, but it's in a more condensed area, so you can hear it in the beach clubs during the day. There's a lot more drinking, I would say, in Greece."—Nicole Kale

- "I was on a barge cruise through Burgundy, France. There was a large emphasis on the food and drink of the area, with multiple wine pairings with all meals and daily visits to wineries."—Mary Chong

- "I went to Vegas and saw the U2 show at the Sphere. I got there two days early. I was meeting a bunch of friends. In my head, it was the first place that

I'd been somewhere unsupervised, by myself, surrounded by alcohol. I was kind of nervous about how I was going to act. I felt uneasy when I got there. Do you remember Pinocchio, when he runs away and he's running through the circus and all those things are popping up in his face? I felt like Pinocchio. I felt like beer was being pushed in my face, or booze. It took me a couple of hours to get comfortable. I don't know if I wanted to drink; I just felt like I was in the wrong place."—Phillip Vitela

- "I really would like to go to Europe. But I probably don't need to go to Oktoberfest."—Maddie

- "I really felt out of place on a cruise I once went on. So much drinking was taking place that at times I felt like my friend and I were the only sober people on the ship."—Anonymous from Texas

- "Spain immediately comes to mind, simply because tapas culture is so strong, and it goes hand in hand with drinking (in the south, you get a free tapa with each drink ordered and it doesn't apply to soft drinks). There's a big culture of wine/beer and I've even watched people leave restaurants in Spain because they don't offer alcohol. Germany has a big beer culture and it almost doesn't 'count' as alcohol, so that's another place where it was very prominent."—Sally, creator of Passports & Plates website

- "My sponsor always told me that wherever I go, to take my sobriety with me. You can go wherever you want; just pay attention to how you really feel and

get some friends you can call anytime, anywhere if
things get weird."—Anonymous from Colorado

• Boozy nightclub shows: "I did actually go to Coco
  Bongo, [a nightlife spot] in Mexico once, and I
  will never do it again. They'll have, like, an Elvis
  or Madonna impersonator and it's really neat when
  they do that. But it starts at 11, first of all, so that's
  hard. And then you pay 75 dollars and it's basically
  all you can drink for free. In the first 15 minutes, the
  waiters and waitresses are walking around pouring
  bottles of alcohol down people's throats. And
  they're getting annihilated. Like the fastest I've ever
  seen people get annihilated before, and I was pretty
  good at that. But that stuff is just so not appealing.
  We were like, 'Can we just have one Diet Coke?'
  They're like, 'Well, you have to have ten at one time.'
  I'm like, 'I'm not going to drink ten Diet Cokes. You
  would be wasting your Diet Coke.' They're like,
  'We only do ten at a time.' It was insane."—Tamar
  Medford, *You're Sober! Now What?* podcaster

• "As a Muslim travel writer regularly talking to Muslim
  travelers, I know for the more conservative Muslim
  [the presence of alcohol] affects their decisions
  about where they would or wouldn't go or reside.
  That said, I am also put off traveling to certain
  regions or 'hotspots' renowned for their drink
  culture."—*anonymous* London-based writer

In general, sober coach Sarah O'Brien advises people newer
in sobriety to avoid big party places. "Places that are maybe
a little bit more family-friendly and not so much bar, club,
casino scene," she suggested. "Not going to Cancun on spring

break, but going to Aruba and horseback riding or riding sand buggies." But you don't have to rule out your dream destinations just because people might be drinking. "You think things like, 'I could never go to Ireland,' or 'I could never go to Jamaica.' But that's not the case. It's just planning events that are fun and filling your time and not planning them around these really crazy parties, late nights, late morning type of deal."

THE EXPECTATION THAT TOURISTS DRINK

Even when you're traveling in countries where there's little or no drinking culture, local people might assume that tourists will drink. And they might want you to, for economic reasons. Drinkers consume more and tip better than those of us who order a soda or a mocktail.

When Phillip Vitela was in Southeast Asia, he saw a lot of foreigners drinking. "Russians, Irish, South Africans, whatever. They're all there. And they're going to drink until their eyes roll back in their head. Southeast Asians aren't big drinkers, but they're going to cater to you. They sell beer on the street. If you're just sitting on a bench, reading a book, they're going to come up to you with some beers. That would be a really difficult place if you were newly in sobriety."

Phillip was still drinking when he visited Southeast Asia, so he was in the middle of the bar scene. I spent almost a month in Cambodia after I'd been sober for 20 years. And while I remember what he was describing, I was able to tune it out and focus on the reasons I was there—as a visiting yoga teacher, and to learn about the country's culture and history.

I really noticed the dichotomy of foreigners versus residents when I visited the Maldives—a 100-percent Muslim country that balances anything-goes resort islands with conservative, traditional non-resort islands that people call "local islands." I was staying at a beautiful resort operated by Maldivian chain Sun Siyam when I took a boat tour to a local island called Guraidhoo. A man named Ali Shaheem was assigned to be my personal tour guide. Since it was a one-on-one tour, I was able to ask all my nosy questions and Ali answered me pretty freely. We saw what local islanders call a "bikini beach," a fenced off area where foreigners can wear their skimpy clothes while lounging on beach chairs or bathing in the sea. Alcohol sales are banned on local islands, but Ali pointed out a Spanish-owned restaurant. "Them guys, they always selling alcohol, but secretly," he told me in a hushed tone. "Normally many, many Europeans need alcohol. Always the nighttime." Yeesh, I thought, what a reputation foreigners have around here.

Sober trip idea: Croatia

Sober traveler Nicole Kale recommends that you skip Croatia's party islands, like Hvar and Pag, which are popular for their array of nightlife options. Visit the smaller islands instead, like Brac, known for its hiking, pristine beaches, and cute fishing villages. I recently spent a week on the sparsely populated island of Prvic for a swimming-themed trip, and found it very beautiful and relaxing.

GROUP TRAVEL WITH SOBER SPECIALISTS

A couple of travel agencies have been organizing special trips for people in recovery for decades. But since the pandemic, a whole new crop of sober tour operators has sprung up. Join one of their tours and you'll find other sober people who can form an excellent support system and take alcohol out of the equation.

Steve A. had been a lifelong traveler when he got sober in 1980. A couple of years into sobriety, he was on vacation at a Club Med. Something about the resort's friendly attitude made him feel comfortable. "All of a sudden I'm waterskiing for the first time sober, and I'm playing tennis for the first time sober. I had previously done nothing sober, and I wouldn't do anything unless I could do it Olympic-caliber for the first time because I have a very low embarrassment threshold." He started to think how great it would be for sober people to go on a vacation like this together. Twelve-step meetings are important, he said, "but the meetings should be fueling and inspiring a bigger life."

That was the seed that grew into Steve's company Sober Vacations International. Ever since the 1980s, he's been a self-described "sober guy who sponsors trips for sober people." He's led and organized sober groups on cruises, golf trips, ski trips, and safaris, and has temporarily taken over resorts, turning them into sober villages. "We've developed a tremendous loyalty," he told me in a 2022 phone interview. "I mean, we left this year and came home with over 400 signups for the next big sober village. These are people who love sobriety and love a bigger life and want to have fun without getting loaded." More than 20 people who met on his sober trips later got married. In addition to fun activities, SVI always has 12-step meetings going on and chances to hear inspirational speakers.

Steve encourages participants to try new things, even when they worry they won't be any good at them. "As my sponsor would say, you need to be willing to be embarrassed to have a bigger life," he said. One year SVI set a Club Med record, with more than 40 of their participants getting certified to scuba dive during a sober village week.

Newer sober travel ventures include Los Angeles-based Choose Life Sober Adventures and Hooked Alcohol-Free Travel, headquartered in Vancouver, British Columbia. Their founders, Cole Bressler and Darci Murray respectively, both had backgrounds in professional group travel before they got sober.

So far, Choose Life offers an active trip to Costa Rica and a cultural trip to Machu Picchu. I went on both trips and found a very nice mix of yoga, outdoors activities, and good food, with a nightly sharing circle or recovery meeting. We all got to know each other well, quickly, and were super supportive. We encouraged the person who was brave enough to zipline despite her fear of heights, and people were nice about my terror of sliding down muddy rainforest slopes and breaking my neck. "People want to live a life," Bressler told me, sounding much like Steve. "They didn't get sober for their life to stay small. They want to do big things and see the world." Bressler has since sold the business to another young sober travel entrepreneur.

David Smith had only been sober a couple of months when he found out about Choose Life. "I really wanted to go on vacation. But I was nervous about traveling because I was so early in my sobriety." He signed up for Choose Life's Costa Rica trip—the same trip I was on. "They have made sure that we have activities all day long," he told me as we sat on my hotel balcony with a view of the pool, a giant iguana sunning nearby. "I can't even put it into words. I have been telling my friends it's the trip of a lifetime for me. I'm doing things I never thought I would do. I'm having more fun as a sober traveler than I ever thought was possible. It's really opened my eyes to the possibilities and the potentials for having fun and enjoying life without having to have a drink."

Hooked trips are less recovery-focused and more about catering to anybody who wants to spend time traveling sans alcohol. Murray is big on full-sensory experiences, whether that be a cold-water plunge or a traditional ice massage in Iceland or a five-star dinner in Morocco. "When you're drinking, your senses are muted. You wake up, you can't see anything. Colors are dimmed. So basically, Hooked is full-sensory travel. You're in touch with all your senses. Colors are brighter, touch is more intense, food tastes better." Her company is still new, but she's aiming to do four to six trips per year.

Isabel van Zuilen leads Sobertopia Retreats in southern Spain for sober-curious women. She promises "sun, yoga, hikes, amazing food, fun activities, relaxation, and connection." And she delivers. I participated in a three-day retreat in Malaga where we stayed in a beachfront villa with both a personal chef and a yoga teacher/massage therapist in residence.

In 2025, EF Ultimate Break, a group travel brand aimed at 18-35-year-olds, added nine new sober tours. "It's very much a business that's powered by Gen Z," EF Ultimate Break president Heather Leisman told me in a phone interview. "So many of our employees fall into the age group of our travelers." When she realized that welcome drinks were going unconsumed and happy hours were under-attended, she began to design alcohol-free itineraries. While some customers are in recovery, many prefer not to drink—and to travel with others who are more focused on adventure and cultural immersion than late nights in the bar.

More sober trips seem to be popping up every time I Google. Recovery Elevator occasionally offers sober trips. Camp Sober Fest, Sobriety Sisterhood, Bigger Life Adventures, Flash Pack, Antigua 12 Sober Travel Retreats in Guatemala, and Your Sober Pal offer trips ranging from inward-looking retreats to group fun around the world. Utah-based Holiday River Expeditions leads affinity group trips, such as a raft trip for LGBTQ+ sober folks. My contact at the company said they're happy to work with a group of sober people to arrange private trips. Sober influencers on Instagram are also leading more and more trips.

All these group trips cost money—maybe more than is in your budget. Chris Marshall, owner of Sans Bar, doesn't want anybody to miss out. "I hope as we move forward that everyone can find experiences that can fit their budget. I don't want to see this be exclusive for just a few people. Because then it's no fun. " Chris organized a few sober camping trips he called Fireside Connections. He reserved three nights of camping space outside of Austin, where he's based. "I would provide the tents and everything that they need. They just need to bring, like, a sleeping bag or a yoga mat. That was like a hundred bucks for the weekend. Those kinds of experiences are things that we need more of. Because again, everyone needs to have these opportunities to do something, to have all the benefits that travel offers."

Sober trip idea: Costa Rica

The best group trip I've been on to date was my adventure trip to Costa Rica with Choose Life Sober Adventures. Our trip had an insane amount of activities—ziplining, river tubing, swimming beneath waterfalls, hiking in the pouring rain, sea kayaking, wildlife watching, standup paddle boarding, and more—plus daily yoga and recovery meetings. We spent three days in the tropical dry forest (which was actually quite wet) near a volcano, and three at the beach on the Pacific coast. The activities were exciting, the lodging very nice, and the company spectacular. Because we were all in recovery, there was no posturing, just support. Everybody was open about their fears and struggles, and we were all laughing together in no time. And even though all the activities were optional, almost everybody participated in everything because nobody wanted to miss out on the nonstop fun.

SOBER CRUISES

*M*any people love cruising. There's the convenience of not having to pack and unpack every day like when you travel by land, the shipboard entertainment, the feel of the water beneath you, endlessly changing views, giant buffets, trays of drinks…wait a minute. Yeah, that last one has been the catalyst for many a slip, and the source of much anxiety. Should we cruise? Is it wise for teetotalers to trap ourselves on a ship full of drinks and drinkers?

People in recovery who want to cruise can try one of the sober travel operators, like Sober Vacations International or Sober Celebrations. Both companies reserve blocks of rooms on large cruises and arrange a solid group of nondrinking cruisers—your support system at sea! You'll still be on a ship with drinkers, but you will have a large sober posse to eat your meals with, explore onshore, and hang out onboard. You could be exploring European Christmas markets with Sober Celebrations, or cruising down the Mekong in Vietnam with SVI.

"When we're in port we recommend staying together for the excursion," Steve A., owner of SVI, explained. "We always reserve a room and have at least one meeting a day so we can connect with people. And dine together. We reserve tables and people stay and engage." They've had up to 500 sober people on a big ship, and 25 or 30 on a smaller cruise to the Galapagos.

Kyle Valenta, an editor at the website Cruise Critic who's been sober for ten years, gave me tips on how nondrinkers can best enjoy cruising. "Cruising has a stigma attached to it, that it's just a giant rager at sea, right?" he said. "Like people are getting really drunk, or they're always going to booze-filled beaches. And that's certainly out there. But stuff has really changed. Staying sober at sea these days is a lot easier thanks to changes in onboard programming, robust mocktail options, and way better entertainment than 10, 20 years ago." The challenges of cruising sober, he said, are similar to those of going to an all-inclusive resort on land.

The first step towards success is doing your research. Valenta stresses the importance of reading customer reviews on Cruise Critic, Trip Advisor, and other crowd-sourced sites. Once you have an idea of what ships and itineraries appeal to you, plug that info into these forums. "You're going to get a sense of, like, is it a party ship? Do people stay up until four in the morning at the nightclub and in all the bars? Or is it a more relaxed ship? Are people talking about how nice and calm it was?" Once you get a feel for the ship's personality, "You can match that against your own tolerances and your own ability to maintain your sobriety in spaces like that on land."

Cruises roughly break down into three categories: Mainstream, expedition, and luxury. The mainstream lines operate large ships and are much more affordable. Expedition cruises focus on nature, active excursions, and education. Many expedition ships carry fewer than 200 passengers, with the largest serving about 500. Luxury cruises can be small or large, and may overlap with expedition cruises.

Each category has its pros and cons. "A smaller ship is going to have a more intimate vibe," Valenta said. "They tend to be much calmer. But you might not always have the wealth of spaces to retreat to, should you want time away in a public setting from a bar or something like that." They're also pricey. I've been on several small expedition cruises and loved them. On a 22-passenger Ecoventura cruise in the Galapagos, the nightly entertainment was a naturalist talking about tortoises, marine iguanas, or geology. However, since the ship was so small, lectures were right by the bar. If this will trigger you, you might want a larger ship.

"Big ships are where you find the drink packages," Valenta said, alluding to various all-inclusive packages passengers buy on some cruise lines. "So you will see a bit more of that lively party vibe generally across the board. However, you're on a big ship. There's a ton of different places and different features onboard that you can use to supplant the times when others are going to drink." Depending on the ship, you might retreat to a gym, library, sun deck or other quiet space. Some ships feature running tracks. Complete 2.4 laps of Royal Caribbean's Wonder of the Seas' two-lane track and you've gone a mile. Valenta likes to work out in onboard gyms in the evening. "In that happy hour timeframe, like 4:30 to 7:30, the gym is empty."

Before committing to a cruise, research its shore excursions. "There are destinations that are very alcohol-centric," Valenta said. "Like a lot of the rivers in Europe that go through wine country or these really historic beer-producing regions. Certainly going to the Caribbean and Mexico there's a type of drinking mind when we go to those places. Puerto Rico, Jamaica, a lot of those islands are really famous for their local rums, and have a craft rum scene. So you will find there are alcohol-centric shore excursions on many cruises, if not most."

Even in the Galapagos, where there are more iguanas than humans, my shipmates and I visited a moonshine operation. I know, it sounds like a minefield for sober folks. But there are alternatives. Valenta recommends taking a history walking tour while on a Mediterranean cruise, or joining active outings anywhere. Cycling, kayaking, and especially snorkeling don't lend themselves to simultaneous drinking. Nor will guides responsible for your welfare be plying you with umbrella drinks before you jump in the water to look at fish.

Expedition cruises have fewer alcohol-centric shore excursions. When I went on a 7-day Alaska voyage with UnCruise, our shore excursions were sea kayaking, bushwhacking, shore walks and wildlife-spotting via an inflatable Zodiac boat. We didn't even visit towns, let alone distilleries. Same when I went on a small cruise to the Norwegian Arctic. Except for a champagne toast on the farthest north piece of ice we could stand on, alcohol was confined to the boat. And even for that toast, staff remembered to bring ginger ale for the nondrinkers. I've found that the smaller the boat, the faster the staff catches on that I don't drink, and they stop offering. Plus, no remotely normal person pays the money and takes the time to go somewhere as hard to reach as the Arctic just to get wasted. These are folks who want to see polar bears. Wellness-themed cruises are another place where heavy drinking is unlikely.

Many of the larger cruise ships have 12-step meetings onboard. You might see one listed on the schedule as "Friends of Bill W. meeting." I was beyond thrilled to find this the first time I ever went on a cruise. My whole family was cruising through Alaska to celebrate my parents' fiftieth wedding anniversary. While they were never heavy drinkers, they enjoyed their pre-dinner gimlet. Perfect timing, as that's when the catchall 12-step meeting was. People showed up with alcohol, drug, food, and gambling addictions. Except for the drugs (as far as I know), the other three were in abundant supply on our ship. If you attend a regular 12-step meeting, you're bound to hear about people's troubles and all the folks they're in conflict with. But onboard, you'd also run into them around the ship. So we had not only the support of seeing each other around, but the added fun of being in the know: oh, that's the irritating sister-in-law, or the brother who won't admit he's a drunk, too.

Internet at sea has improved greatly, so if there's no onboard 12-step meeting, or you prefer staying connected with your home group, the connection is often sufficient to attend virtual meetings.

Cruise lines have been amping up their mocktail menus, so you don't have to suffer from umbrella drink envy. Valenta raves about Virgin Voyages' mocktail program. "Virgin has a really wonderful Mexican restaurant. They're doing mocktails with flavors that would make sense with Mexican food. You're not just getting a replication of, like, a sober margarita. You're getting something a little more interesting." Luxury line Silversea does craft mocktail pairings, and the cruise yacht brand Explora is also making a name for itself with mocktail enthusiasts. On the bigger lines like Carnival, Royal Caribbean, and Norwegian, you can get non-alcoholic beverage packages that will include your mocktails. Valenta felt like he blended in with other passengers as he sipped virgin piña coladas on a Royal Caribbean cruise.

As cruise lines compete against each other for passengers, the entertainment has changed, too. On the bigger ships, shows keep getting more spectacular. Valenta was captivated by a water show on a Royal Caribbean cruise. "People are leaping from, I don't even know, five decks above into pools that look not remotely deep enough. It's just incredible. It's a very different type of energy than sitting in a theater. You don't feel like you need to add to the experience."

If you and your sober friends are really feeling rich, some cruise lines with smaller vessels, such as Lindblad Expeditions, let you charter the whole ship.

NONDRINKING GUESTHOUSES AND RESORTS

*A*round the world, a few nondrinking innkeepers run dry guesthouses. For teetotalers like me, the less readily available alcohol is, the more relaxed I feel. Some retreat centers also have alcohol-free campuses to promote physical health and mental clarity. One challenge is that these can be hard to track down, and there's always the chance that they will cave and start serving alcohol. I went to visit a high-end dry hotel in Florida recently, only to be handed a cocktail menu. The week before they'd started serving in an understandable attempt to stay in business. Still, some places manage to remain open without serving alcohol. Just double check when you're ready to book to make sure policies haven't changed.

Sober properties pop up in random-seeming places, usually the work of nondrinkers. On Big Corn Island in Nicaragua, you can enjoy beaches, hammocks, and a pool with a swim-up mocktail bar! Sea Star Spa Dry Resort features four cabanas and the mission "to provide a tropical paradise sanctuary for people that desire a fun beach vacation without the presence of alcohol." On the Isle of Wight, the Somerton Lodge Alcohol Free Hotel also promises a quiet break from hustle and bustle—and booze.

I've found promising lists of alcohol-free properties on halal sites designed for Muslim travelers. Going to Turkiye, the Maldives, or Morocco? The halal sites have you covered. Please dress modestly and be respectful if you visit these. Sally, a Muslim travel writer whose site Passport & Plates uses the brilliant tagline "Travel and dine sans spirits and swine," recommends Malaysia, Indonesia, Singapore, Turkiye, Morocco, Egypt, the UAE, and Qatar as places where you can find rich nondrinking cultures.

> ### Sober trip idea: Marfa, Texas
>
> Alysse Bryson, founder of The Sober Curator empire, recommends this eclectic little town in the desert, famous as the set of *Giant*, James Dean's last film. Nowadays art galleries, boutiques, and unusual lodgings like teepees and yurts fill the town. Alysse recommends signing up (in advance, they sell out) for a star party at the Rebecca Gale Telescope Park. Or go out on your own late at night to watch for the mysterious Marfa Lights, which may be caused by ghosts, UFOs, distant car headlights, or electric energy.

WELLNESS TRAVEL CAN BE YOUR FRIEND

*D*epending on the person, a week of meditating, eating health food, and doing lots of yoga could be paradise or punishment. For example, Absolute Sanctuary in Thailand is alcohol-free, but you'll also be expected to fast, give up coffee, and partake in colon cleanses. Some folks might say, "That's a vacation?!" But if wellness travel appeals to you, you'll find a world where alcohol use is minimal and the focus is squarely on improving your life.

The Global Wellness Institute (GWI), a nonprofit research organization, defines wellness tourism as "travel associated with the pursuit of maintaining or enhancing one's personal wellbeing." It sees travel as a way to combat rampant unwellness. GWI also notes, "Wellness tourism is the powerful intersection of two large and growing industries: the $2.6 trillion tourism industry and the $4.5 trillion wellness industry (2017 figures)." So don't expect a wellness vacation to be as free as all that air you'll be breathing in your meditations.

Do some research first. I've been on wellness vacations with friendly groups of people from all walks of life who were spending a few days doing lots of yoga or meditation while staying in a comfortable hotel. These felt like real vacations. Sometimes you'll encounter very earnest people in hundred-dollar yoga pants who can talk for half an hour about thoroughly chewing their food to extract all of its prana. This feels less like vacation. Once I even found myself under a giant golden meditation dome in the backwoods of Tennessee, being shown around the complex by a woman who'd devoted 20 years to her guru—without pay. To my American mind, at least, that screamed cult and made me uncomfortable. Instead of concentrating on my breathing, I worried about whether the guru would take care of her if she ever got to retire.

In Breckinridge, Colorado—known as a party/ski town—Katherine Grimm owns Clairvergence Wellness Center. She offers wellness retreats that include kundalini yoga, workshops about habit change, and acudetox, a special kind of acupuncture that seeks to reduce cravings by targeting five points on the ear. While her retreats aren't exclusively for sober travelers, these events are alcohol-free. "They are designed to help us find our more authentic self," she told me in a phone conversation.

Wellness doesn't have to be expensive. I naturally gravitate toward things like yoga, vegan food, nature, and exercise, so I almost always work some wellness components into every trip. When I travel, I usually bring a yoga mat, workout clothes, and running shoes. I try to find a hotel with a decent gym. Sometimes I check out local yoga classes if there's a studio nearby, get a day pass at a gym, or do a drop-in class at a YMCA.

DESTINATION SPAS

If you have the money and want to have a luxurious and quiet time, consider visiting a destination spa. There are lots of these around the world, and magazines like *Travel + Leisure* are always printing best-of lists. Here are a few that I've either visited or heard good things about as far as an alcohol-free reset.

- **Hilton Head Health in South Carolina** is an island resort with a long history of getting people on track with healthy habits. Behavioral psychologist Peter Miller founded Hilton Head Health in 1976 in conjunction with a regional hospital as an alternative to the popular "fat camps" of the day. Expect a schedule full of exercise classes, motivational talks, healthy meals, and cooking demos. They also offer activities that feel less regimented, like bike riding, paddleboarding, and kayaking. The setting is beautiful—a jungly campus full of Spanish moss with beach access.

- In the Berkshire Mountains of Massacushetts, **Kripalu Center for Yoga & Health** offers workshops focusing on topics in the areas of yoga, creativity, spirituality, health and self-discovery. Or you can do an individual retreat and choose whether to attend yoga classes and self-improvement lectures or just wander the property's 100 acres of forest and private lakefront. I visited in winter and participated in a snowshoeing and mindfulness workshop. The grounds were beautiful, the vegetarian food was delicious, and Kripalu's bustling dining hall does not serve alcohol, neatly taking that temptation or annoyance out of the equation. A trip to Kripalu is recommended for nonsmokers who love nature, enjoy vegetarian food and yoga, and have a high tolerance for earnest people on various spiritual paths. Lodging options range from the money-saving dorm rooms with shared baths to private rooms with private baths.

• At the **Art of Living Retreat Center** in the Blue Ridge Mountains of rural Boone, North Carolina, you can do an individual R&R retreat, or sign on for a focused program like a multi-day workshop on developing intuition or extending your longevity. Expect lots of yoga and meditation, hiking trails, a clay studio, vegetarian food, and no alcohol served (although people are allowed to bring their own and consume it in their rooms). Sometimes there are recovery-specific courses offered, so check the center's calendar.

• The grand **Fairmont Chateau Lake Louise**, built in 1890, is in Banff, Canada's first national park. It sits right at the edge of Lake Louise, which is an amazing blue in the warmer months. The hotel offers wellness retreats centering on activities like mindfulness, yoga, and meditation. It's a good mix of luxury hotel, scheduled activities, nature, self-improvement, and a built-in group of buddies. When I got there for a mindfulness retreat in April, I was like, 'Where's the famous lake?' Underneath a lot of snow and ice, as it turned out. I had time for some lovely snow hikes. This isn't strictly an alcohol-free trip—people will still be tippling in the hotel bar—but it's not chic among the wellness crowd to hit the bottle hard.

- If you want to make like a real yogi and retreat into the Himalaya Mountains, you could travel to **Ananda Himalaya Spa in Rishikesh, India.** Top travel publications have endorsed it as the world's best spa, and Oprah Winfrey also gives it rave reviews. If you yearn to go somewhere extremely beautiful where your health is top of everybody's mind, this palatial destination spa in the famously spiritual city of Rishikesh might be for you. While alcohol is available, it's not encouraged. The spa's senior physician has spoken about the negative effects of drinking, and the benefits of taking a break from alcohol or quitting entirely. At Ananda, the staff encourages visitors to make better health decisions with yoga, massages, emotional healing therapies, and a seemingly endless selection of spa modalities.

OH NO, I ACCIDENTALLY DRANK ALCOHOL!

I try to be vigilant, but alcohol has still snuck into my mouth a few times. Once I was at a nice restaurant with some other travel writers while on a press trip in Quebec. I ordered sparkling apple juice and the server accidently brought me hard cider. I thought it tasted like alcohol so I quietly backwashed it into the bottle. Then I said to my companions, who knew by then that I didn't drink, "I think this has alcohol in it." Immediately the guy next to me grabbed the bottle and took a sip. "Yep," he said. He passed it around and everybody confirmed. I was in a mild state of shock with two simultaneous thoughts: *Oh my God, I just had alcohol in my mouth,* and *Oh my God, everybody's drinking my backwash.*

Freelance journalist Fred Wright also had an unwanted brush with booze while traveling. "I discovered on a press trip to the Smoky Mountains of Tennessee that 'drinking' can include infused foods such as cherries infused with moonshine. I hadn't considered that as a risk during a food tasting."

While I'm not a very religious Catholic, I do enjoy going to masses in beautiful old churches, especially when traveling. One Sunday morning in Detroit, I went to a record-setting (for me) three masses in a row. The third was at Holy Family, an Italian church. It was so old-fashioned that instead of putting communion in your hands, the priest insisted on putting it directly in people's mouths. At every mass I'd ever been to, people would accept the bread (body of Christ) and then they could take a sip of wine (blood of Christ), or not. I always skipped the blood. But once the priest put the communion in my mouth, I realized they'd devised an all-in-one system of bread pre-dipped in wine.

Communion is considered super sacred in the Catholic church. If anybody saw me spit it in a trashcan, that priest probably would have personally sent me straight to hell. So I swallowed it. And tasted wine the rest of the mass. As soon as it was over, I drank coffee to get the taste out of my mouth and called my husband, who is also a nondrinker. It wasn't like accidentally getting the taste of wine or hard cider unleashed me on a deadly drinking rampage, but it was unsettling and made me very uncomfortable and not so happy.

What if you unexpectedly find alcohol in your mouth? "Don't panic," said sober coach Sarah O'Brien. "I think that this happens." Several times she's ordered a soda when she's out with her cousins and been given the wrong drink. "And I'll say to one of my cousins, 'Can you taste this, does this have something in it?' It doesn't really panic me anymore. Because I recognize this isn't for me. And I'd put it down." She encourages people in recovery to work their sobriety program so that they're not held back from experiencing things in life. "The work that you can do in sobriety is super crucial."

Hospitality professional Makina Labrecque emphasizes that accidentally drinking doesn't mean you've relapsed. "I've had friends who were nondrinkers who have been served alcohol.

That doesn't mitigate your sobriety. That doesn't change your intention." If you accidentally take a sip of alcohol, stay calm and put the drink down. The bartender just made a human error. "You can't assume everyone's going to care as much as you."

Vigilance will help you avoid at least some of these situations. If you're at an event and something looks like punch, ask if there's alcohol in it. Read labels and make sure you're getting a soft drink, not a hard cider. If you're in a noisy bar or restaurant, reconfirm with your server that you're getting a mocktail, not a cocktail.

Three times while traveling in Mexico, sober podcaster Tamar Medford found liquor in drinks she'd ordered as virgin. "Once was tequila. And I'd taken a sip. And I spit it out. But just from having that alcohol, and because tequila's so strong, my whole mouth went numb. And it was like instantly I had tingles all over. You have to be so careful. Because if you sip and swallow, depending on your type of recovery, you could be like, 'Oh, well, I just screwed up my recovery, I might as well have the rest.' Or you're sitting there craving the rest of that drink. But lucky for me, I was with somebody who was sober. And I said, 'Oh my God, this has got alcohol in it.' And I gave it to the waiter. I said, 'Here, I can't have this.' And they apologized profusely. But if you weren't with someone who's sober, that could be a very dangerous situation."

Bring your own non-alcoholic drinks to parties and events. And if you do find alcohol in your mouth, spit it out discreetly, if possible. And get on with your sober life.

OH MY GOD, I RELAPSED! WHAT TO DO ON THE ROAD?

onderful as travel is, it ups the odds of relapsing. "You're totally in a strange environment and you're out of your comfort zone," said Steve A. "What's an alcoholic going to do with anxiety? Pour alcohol on top of it." And depending on what sort of trip you go on, drinks might be very easy to come by. "If you're at a luau in Hawaii, they want you to loosen up. You're sitting on a beach, they're trying to bring you drinks. You know, it's challenging."

Be aware of the difference between a slip and a relapse. A slip is an isolated incident. Relapse is returning to a previous pattern of drinking. Sober traveler Maddie was flying to visit her friend in Los Angeles. "The flight attendant was offering wine," she said. "Which I did drink. It was one of those things where I was like, 'Well, I'm flying alone, who's going to know?'" But she knew. And she felt awful. "Luckily I didn't drink a lot of it. But I got there and I was really upset. My friend has been my best friend since first grade. She was just like, 'This isn't a relapse. This is a slip.' She was really gracious and supported me emotionally. I got there and had this amazing person who just like, loves me. And I didn't love myself very much." What if Maddie hadn't had that support when she got to Los Angeles? Perhaps the slip would have escalated into a relapse.

Sober coach Sarah O'Brien emphasized the importance of support if you relapse on the road. "Reach out to people that you have formed relationships with as soon as you can. Picking up the phone after a relapse can be damn near impossible. The guilt, the shame, the remorse are almost too much to bear. But know that there are people in your corner who can help you get back up. This doesn't need to be the end of the story." She's known people who relapsed while on important business trips. "And maybe the stress or the pressure was just a little bit too much. Unfortunately, it led them to a drink. It's getting back into that recovery-based program that had brought you to a place where you could get this job and have these opportunities. So I think it's very important to understand that if something does happen, there is a way out."

If the relapse is severe enough that you need medical attention, your options will vary. "Depending on where they are, they may be limited, or there may be extensive addiction medicine professionals that may come in to help," said George Koob. "I'm assuming that virtually in any part of the western world, if you really had an alcohol withdrawal event, that they would know the protocols for treating that. Because as you know, alcohol withdrawal can be life-threatening if it's really severe." It's likelier that you'll be scared, ashamed, lonely, and desperately need your support system. Dr. Koob reminds you to keep your cell phone charged so you can call your people.

"If you do drink, don't beat yourself up; if you are an alcoholic it's only natural, and being sober is the unnatural state," an anonymous sober source from Colorado told me. "I think this just means that it's the go-to when things get too intense." While she thinks you should be kind to yourself, you should also take slips and relapses seriously. "If staying sober is important to you then you know that every slip just makes it harder the next time you have to resist that craving. People who aren't alcoholic don't have that craving, by the way. It never gets better unless you stop, for real. Playing games with it will only prolong the suffering."

What about gray-area drinkers, those people who are nowhere near the abyss of alcoholism but who have given up drinking to live their best and healthiest lives? Without the high stakes, does a slip even matter? "Oh, the emotional experience is the same, even though maybe the long-term danger isn't the same," Amanda Kuda, gray-area sober coach, told me. "The emotional feeling of disappointment in oneself is all there."

Amanda works with her clients to get at what caused the slip. "The thing I always have them ask is, 'What was I trying to emotionally achieve or avoid in this situation? Was there a feeling I was trying to shortcut myself to or bypass with this experience?'" Usually they say they just wanted to relax. "But behind that 'I just wanted to relax' was 'I wanted a shortcut to relaxation and I don't know how to do it myself.' So I always have them just kind of evaluate what they were trying to get towards or away from. And then to regroup." Like the other experts I talked to, Amanda urges people to forgive themselves. "Say, 'Tomorrow's a new day. I don't have to go into a downhill spiral because of this one instance.'"

THE ABSTINENCE VIOLATION EFFECT

A phenomenon known as the abstinence violation effect is one reason a slip can easily turn into a relapse. Perhaps you've had the experience of eating half a box of cookies and then saying, "Fuck it, I might as well eat the other half." Or you drank half a bottle of gin and said, "Fuck it, I might as well drink the rest." If so, you know what I'm talking about, and you understand why the abstinence violation effect has also been called the "fuck-it phenomenon."

In a 1994 interview published in the *Journal of the Academy of Nutrition and Dietetics*, the late Dr. Alan Marlatt talked about cigarette smokers trying to quit their habit. He explained, "He or she attributes the lapse to failure of character, e.g., lack of willpower, and motivation collapses. This negative emotional state sets the stage for the individual to smoke the next cigarette."[25] Instead, Marlatt suggested what he called the habit model, where "mistakes and errors are expected as part of the learning process. The individual needs to look at mistakes constructively rather than moralistically, to learn from them instead of giving up or giving in."

However, many of us face the abstinence violation effect (AVE), which leads to both more extreme behavior during relapses, and more pessimism and negativity about their meaning. Twelve-

---

25 "Mindfulness and metaphor in relapse prevention," www.jandonline.org/article/0002-8223(94)92361-2/pdf

step programs, for all their good, contribute to the AVE. AA members celebrate their length of sobriety with plastic or metal tokens which carry a lot of symbolism and importance. Picking up a drink means starting over with a 24-hour chip. Who wants to lose months or years of sobriety and go take a newbie chip in front of all their friends in a meeting?

"From my experience, if I know my own self if I was to pick up a drink, that would probably be my next thought: that I had already done it, so I'm going to go really hard and continue to do it," said Sarah O'Brien. "It doesn't need to be like that, but for a lot of us, myself included, that alcoholic mind, it's like go big or go home. I've already done it. I've lost my sobriety. I have to restart my days. I've done these things. I might as well go hard. That's very, very common."

Slips and relapses are very emotional times. But if we can gain a little perspective, maybe we can try viewing things more like Marlatt. "People tend to over-identify with their problems," he said. "They think 'I'm obese,' 'I'm an alcoholic,' 'I'm an addict,' rather than 'I engage in certain behaviors that have certain negative consequences and these behaviors can be changed.'" Realizing that you have a choice, that you can change, is necessary to avoid the abstinence violation effect.

On relapse

"Don't do it. I think that not drinking and traveling is for me the only way that I get to actually experience the traveling. I only have this one life that I know of. And I want to experience every single moment. Especially when I get to go places. I love getting to go places. I want to experience every single moment so that when I'm older and I'm on my deathbed I can remember all those beautiful places I've been, and the people that I've met. And if I'm not sober while I'm traveling, that is not a possibility."

—Sarah

40 THINGS TO DO ON A TRIP OTHER THAN DRINK

Feel like you're at loose ends and getting anxious? Too much unscheduled time can feel frightening, and sometimes I'm just not sure what to do with myself. It can help me feel more secure to just pick an activity, any nondrinking activity, and have a plan. Eventually the anxiety passes. Check out this list of things to do and just pick one.

1. Go for a walk

2. Take a class at a local yoga studio

3. Visit a park

4. Attend a local 12-step meeting

5. Drink a cup of coffee or tea at the nearest café

6. Find an empty bag and go pick up some trash

7. Meditate for 10 minutes

8. Take a nap

9. Go see a movie at a theater

10. Visit the local historical museum

11. Take a guided tour—bus, bike, or walking

12. Call a friend

13. Take a bath or shower

14. Rent a kayak and go for a paddle

15. Work out in the hotel gym

16. Buy a scoop at the local ice cream shop

17. Ride a local bus for a cheap sightseeing tour

18. Go to a botanic garden

19. Walk through a residential neighborhood and find a
cat to pet

20. Write in your journal

21. Do some sun salutations

22. Listen to music

23. Watch TV

24. Play a game on your phone

25. Sing a song

26. Dance around your hotel room

27. Write a letter to your congressperson

28. Get a manicure

29. Edit photos in your phone

30. Do an online language lesson

31. Ask the hotel clerk for the best attractions you can
walk to that don't involve drinking

32. Write postcards to friends and family

33. Start a craft project

34. Teach yourself something on YouTube: hair braiding, magic tricks, yodeling

35. Google the town's history

36. Write a thank you note to the hotel housekeeper

37. Order room service

38. Set some goals for the next month, year, and five years

39. Clean out your email inbox

40. Try on clothes at local boutiques

"I DON'T DRINK ALCOHOL" IN A BUNCH OF DIFFERENT LANGUAGES

*A*round the world, people speak more than 7,000 languages. That's a lot of ways to say "I don't drink alcohol." While we can't include all of them, here's how to convey your message in 30-plus languages spoken widely by tourists. These are phonetic and may be tricky, so get a local to help you practice the correct pronunciation when you arrive.

Arabic: انا ال أشرب الكحول

(ʻAna la ʻashrab alkuhul)

Chinese: 我不喝酒

(Wǒ bù hējiǔ)

Croatian: Ne pijem alkohol

Czech: Alkohol nepiju

Danish: Jeg drikker ikke alkohol

Dutch: Ik drink geen alcohol

Estonian: Ma ei joo alkoholi

Filipino: Hindi ako umiinom ng alak

Finnish: En juo alkoholia

French: Je ne bois pas d'alcool

German: Ich trinke keinen alkohol

Greek: Δεν πίνω αλκοόλ

   (Den píno alkoól)

Haitian Creole: Mwen pa bwè alkòl

Hindi: मे शराब नहीं पीता हूँ

   (me sharaab naheen peeta hoon)

Icelandic: Ég drekk ekki áfengi

Indonesian/Malay: Saya tidak minum alkohol

Italian: Non bevo alcolici

Japanese: 私はアルコールを飲みません

   (Watashi wa arukōru o nomimasen)

Khmer: ខ្ញុំមិនដឹកស្រាទេ។

   (Khnhom min phoekasra te)

Korean: 나는 술을 마시지 않는다

   (Naneun sul-eul masiji anhneunda)

Nepali: म रक्सी पिउँदिन

   (Ma raksī pi'uṁdina)

Norwegian: Jeg drikker ikke alcohol

Portuguese: Eu não bebo álcool

Russian: я не пью алкоголь

(Ya ne p'yu alkogol')

Sinhala: මම මත්පැත් බොත්තනේ තැහැ

(Mama matpæn bonnē næhæ)

Slovenian: Ne pijem alkohola

Spanish: No bebo alcohol.

Swahili: Sinywi pombe

Swedish: Jag dricker inte alkohol

Tamil: நான் மது அருந்துவதில்லை

(Nāṉ matu aruntuvatillai)

Thai: ฉันไม่ดื่มแอลกอฮอล์

(C̄h̀ạn mì dụ̀m xǣlkxḥxl)

Turkish: Alkol içmem

Vietnamese: Tôi không uống rượu

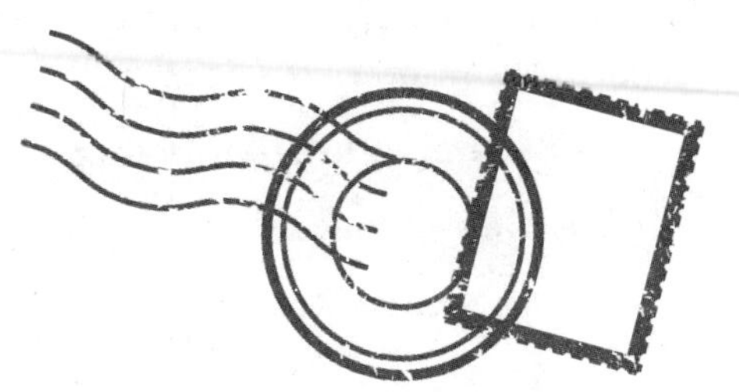

RECOVERY GROUP RESOURCES

Help is just a click away! These sobriety groups meet around the world. And if you're not in the right spot for an in-person meeting, connect virtually.

Alcoholics Anonymous

www.aa.org

Jewish Alcoholics, Chemically Dependent Persons and Significant Others (JACS) and JCS Recovery

jewishboard.org/listing/jacs-jcsrecovery

jewishboard.org/how-we-can-help/jewish-community-services/jcs-recovery

Life Recovery Christian 12-step

liferecoverygroups.com

Lifering Secular Recovery

lifering.org

Millati Islami

www.millatiislami.org

Refuge Recovery Buddhist group

www.refugerecovery.org

SMART Recovery

www.smartrecovery.org

Women for Sobriety

womenforsobriety.org

ACKNOWLEDGEMENTS

I'd been traveling sober for a long time when I started researching this book. And while I didn't know it all, maybe I thought I knew a lot. Well, I quickly learned how many different kinds of sober travelers were out there and that they had countless insights and tactics regarding sober travel that I'd never thought of. So I'm extremely grateful to them for sharing their experience. Some chose to be identified by their real names, some by a pseudonym or just their first name.

Here's my crew of helpers: Steve A., Candy Arrington, Sam Bellantoni, Martin Betch, Cole Bressler, Alysse Bryson, Mary Chong, Tim Cigelske, Brandy Cogsdill, Ariel Dunitz-Johnson, Abby Ehrmann, Katherine Grimm, Jason, Lane Kennedy, Breanne Kiefner, George Koob, Nicole Kale, Amanda Kuda, Kylie, Makina Labrecque, Eliabeth Harryman Lasley, Eric Lucas, Maddie, Philipp M., Lisa Maloney, Chris Marshall, Jeremiah Mason, Dani Mathews, Sarah, Matt McDermott, Andy McMillan, Tamar Medford, David Messerli, Sarah Michaud, Mira T., Darci Murray, Carly Nelson, Héctor Nguema, Alexandra Nyman, Sarah O'Brien, Seth O'Malley, Emma Orr, Mary Paige Rose, Rebecca Rush, Sally, Mike Sanchez, Roger Sanchez, Ali Shaheem, Juliana Sims, David Smith, Jason Steinhauer, Sugar, Kyle Valenta, Isabel van Zuilen, Phillip Vitela, Ernest White II, Fred Wright, and several folks who chose to remain anonymous.

Also, a special shout out to the Society of American Travel Writers Sober Travel Affinity Group. And thanks to my husband, Gideon Parque, our black cat Lucifer, and my family for their support.